Praise for *Let*

"This book might change your perspective on real cleanliness . . . and along the way help you to raise healthier kids."
—Giulia Enders, author of *Gut: The Inside Story of Our Body's Most Underrated Organ*

"[They] make that case with an unusually convincing display of evidence—as well as historical anecdotes and a parent-friendly sense of humor."
—*The Washington Post*

"A must-read for parents, teachers and any healthcare provider for children, *Let Them Eat Dirt* takes you inside the inside tract of a child's gut, and shows you how to give kids the best immune start early in life."
—William Sears, MD, co-author, *The Baby Book*

"I loved this book."
—Dr. Richard Besser, president and CEO, Robert Wood Johnson Foundation

"With the quiet weight of their authority, pioneer researchers Brett Finlay and Claire Arrieta help parents to understand the real nature of microbes, and then to act to improve their children's health."
—Martin Blaser, author of *Missing Microbes* and director of the NYU Human Microbiome Program

"As a parent and a microbiologist, I appreciated the up-to-date and actionable science that *Let Them Eat Dirt* highlights, including the groundbreaking work conducted in the authors' own lab. As a Professor of Pediatrics, I appreciated the accessible format and writing style that makes this wealth of information and its limitations easy to understand for the increasing crowd of parents who are concerned about their children and their growing microbiomes. *Let Them Eat Dirt* gives an entertaining, engaging and accurate view of what we're discovering about the microbiome and why it matters for you and your children."

—Rob Knight, professor of pediatrics and computer science & engineering, and director, Center for Microbiome Innovation, UC San Diego, and author of *Follow Your Gut: The Enormous Impact of Tiny Microbes*

"What a triumph. This book should be read by every pregnant woman, every parent, every pediatrician. It's not just a great read but terribly important."

—Professor Margaret McFall-Ngai, member of the National Academy of Sciences and director of Pacific Biosciences Research Center, University of Hawaii

"Great book! Very clear, down to earth, and interesting; it reads like a story! *Let Them Eat Dirt* takes an important and complex subject and makes it less scary."

—Eran Elinav, MD, PhD, principal investigator, Host-Microbiome Interaction Research Group at the Weizmann Institute of Science, and senior fellow, Canadian Institute for Advanced Research

"Solid, easily assimilated evidence showing how microbes are an integral part of a child's healthy life."

—*Kirkus Reviews*

"Finlay and Arrieta explain, in illuminating detail, the importance of the gut microbiome . . . They hope to restore the powerful benefits of microbe transfer from the environment to the young child, benefits lost as a side effect of efforts to reduce infectious disease risk and of cultural attitudes that conflate dirt with disease . . . The focus on practical choices before and during birth makes this book a good resource for expectant parents."

—*Publishers Weekly*

Let Them Eat Dirt

Let Them Eat Dirt

How Microbes Can Make Your Child Healthier

B. BRETT FINLAY, PhD

MARIE-CLAIRE ARRIETA, PhD

ALGONQUIN BOOKS OF CHAPEL HILL 2017

Published by
Algonquin Books of Chapel Hill
Post Office Box 2225
Chapel Hill, North Carolina 27515-2225

a division of
Workman Publishing
225 Varick Street
New York, New York 10014

First paperback edition, September 2017. Originally published
in hardcover by Algonquin Books of Chapel Hill in 2016.
Printed in the United States of America.
Design by Steve Godwin.

The Library of Congress has catalogued the hardcover edition as follows:
Names: Finlay, B. Brett, [date] author. |
Arrieta, Marie-Claire, author.
Title: Let them eat dirt : saving your child from an oversanitized
world / B. Brett Finlay, PhD, and Marie-Claire Arrieta, PhD.
Description: Chapel Hill, North Carolina : Algonquin Books
of Chapel Hill, 2016. | Includes bibliographical references.
Identifiers: LCCN 2016018794 | ISBN 9781616206499 (HC)
Subjects: LCSH: Medical microbiology. | Bacteria—Health
aspects. | Pathogenic microorganisms.
Classification: LCC QR46.F56 2016 | DDC 616.9/041—dc23
LC record available at https://lccn.loc.gov/2016018794

ISBN 9781616207380 (PB)

10 9 8 7 6 5 4 3 2 1
First Paperback Edition

To our kids, Jessica, Liam, Marisol, and Emiliano,
for inspiring us to get the word out that kids need
more dirt in their lives.

Contents

Preface

We all want what is best for our kids. The problem is that there is no perfect handbook on how to raise them, nor is there any one best way, either. We read books and articles, talk to friends, and try to remember (or forget!) how our parents raised us. Both of us have children and have struggled and muddled through the parenting process the same way everyone does. We are also scientists who have worked with microbes for many years, and we couldn't help but consider how these ever-present microbes influence development as we raised our children. At first we studied microbes that cause disease, and we feared them just like anyone else. But more recently we began taking notice of all the other microbes that live in and on us—our "microbiota." As we continue to study the microbiota of humans, it is becoming clear that our exposure to microbes is most important when we're kids. At the same time, modern lifestyles have made childhood much cleaner than ever before in human history, and this is taking a huge toll on our microbiota—and our lifelong health.

The genesis of this book came from the realization that the studies in our lab—and the labs of several other researchers—prove that microbes really do impact a child's health. What shocked us most was how early this starts—the first one hundred days of life are critical. We knew microbes played a role in well-being, but we had no idea how soon this role began.

Several other factors converged to help convince us to write this book. Claire has young children, and all of her young parent friends were extremely interested in the concept of microbes and how they might affect their kids. Whenever we tell other parents about our work, the questions never cease—*Do I need to sterilize their bottles every time? What kind of soap should I use?* We realized that there are many questions out there about microbes . . . and a lot of wrong information.

Brett is married to a pediatric infectious disease specialist (Jane) who was constantly suggesting articles and findings about how microbes affect kids, which led us to realize that since this was such a new field, there was no one source parents could turn to if they wanted to learn more. Not to mention that scientific articles are usually dry, terse things with lots of jargon and, frankly, are terribly boring. However, this new area of research has a lot to offer to people raising children who are not likely to get this important information from dense scientific papers or from studies often misinterpreted by the press. There is a lot of information being produced by some of the best scientists in the world, which we consider extremely useful for the day-to-day decisions we make while raising our children, so we felt compelled to gather it all in one book and make it accessible to the everyday parent.

We start off by explaining a bit about microbes, and then explore what happens to a pregnant woman's body in terms of her microbiota and how it affects her child(ren) for life. We then discuss the delivery process, breastfeeding, solid foods, and the first years of life from a microbial perspective. In the middle of the book we cover lifestyle issues (*Should I get a pet? What do I do with a dropped pacifier?*) and the use of antibiotics. The latter part of the book features chapters dealing with specific diseases that are growing by leaps and bounds in our society, and the microbes that seem to affect them. These include obesity, asthma, diabetes, intestinal diseases, behavioral and mental health disorders such as autism, and a whole array of diseases in which, even five years ago, we had no clue microbes might be involved. Readers may want to skip over particular chapters if you feel that they are not applicable to you. However, each one is full of information that will educate you about the processes involved in these health issues. We think the section on the gut–brain connection (chapter 14) is particularly interesting in its exploration of how microbes might affect the brain and mental disorders. We finish the book with a discussion on vaccines and a futuristic view of what we can expect in terms of new therapies and medical interventions in the next few years. Each chapter ends with a few Dos and Don'ts—these are not meant to be comprehensive medical advice, but suggestions about things to do (or not do) that are based on current scientific evidence.

What we have learned in writing this book, and what we hope to convince readers of, is that microbes play a very large part in our children's lives. Even as scientists in the field, we were stunned to discover some of the profound roles these microscopic bugs have in normal childhood development. No doubt many of these findings, and many more to come, will have a major impact on how we think about raising our children.

—B. Brett Finlay and Marie-Claire Arrieta

Let Them Eat Dirt

PART ONE

We Are More Microbe Than Human

1: Children Are
   Microbe Magnets

Microbes: Kill Them All!

Microbes are the smallest forms of life on Earth. They encompass bacteria, viruses, protozoa, and other types of organisms that can be seen only with a microscope. Microbes are also the oldest and most successful forms of life on our planet, having evolved long before plants and animals (plants and animals actually evolved from bacteria). Although invisible to the naked eye, they play a major role in life on Earth. There are an astounding 5×10^{30} (that's 5 followed by 30 zeroes!) bacteria on Earth (for comparison, there are "only" 7×10^{21} stars in the universe). Collectively, these microbes weigh more than all the plants and animals on the entire planet combined. They can live in the harshest and most inhospitable environments, from the Dry Valleys of Antarctica to the boiling hydrothermal vents on the seafloor—they can even thrive in radioactive waste. Every form of life on Earth is covered in microbes in a complex yet usually harmonious relationship, making germophobia the most futile of phobias. Unless you live in a sterile bubble without any contact with the outside world (which is a time-limited proposition; see Bubble Boy, page 15), there is no escaping microbial life—we live in a world coated in a veneer of microbes. For every single human cell in our bodies, there are ten bacterial cells inhabiting us; for every gene in our cells, there are one-hundred fifty bacterial genes, begging the question: Do they inhabit us or is it really the other way around?

While in its mother's womb a baby is for the most part sterile, but at the moment of birth it receives a big load of microbes, mainly from its mother—a precious first birthday gift! Within seconds, the baby is covered in microbes from the very first surfaces it touches. Babies born vaginally encounter vaginal and fecal microbes, whereas babies born via C-section pick up microbes from the maternal skin instead. Similarly, babies born at home are exposed to very different microbes than if they are born in hospitals, and different homes (and hospitals) have different microbes present.

Why does all this matter? Well, until very recently hardly anyone thought it did. Until recently, whenever we thought of microbes— especially around babies—we considered them only as potential threats and were concerned with getting rid of them, and it's no surprise why. In the past century, we have experienced the benefits of medical advances that have reduced the number and the degree of infections we suffer throughout life. These advances include anti-biotics, antivirals, vaccinations, chlorinated water, pasteurization, sterilization, pathogen-free food, and even good old-fashioned hand-washing. The quest of the past hundred years has been to get rid of microbes—the saying was "the only good microbe is a dead one."

This strategy served us remarkably well; nowadays, dying from a microbial infection is a very rare event in developed countries, whereas only a hundred years ago, seventy-five million people died worldwide over a span of two years from the H1N1 influenza virus, also known as the Spanish flu. We have become so efficient at avoiding infections that the appearance of a dangerous strain of *Escherichia coli* (aka *E. coli*) in a beef shipment or *Listeria monocytogenes* in spinach leads to massive recalls and exportation bans, along with accompanying media hysteria. Microbes scare all of us, and rightly so since some of them are truly dangerous. As a result, with very few controlled exceptions such as yogurt or beer, we often think that the presence of microbes in something renders it undesirable for human use. The word antimicrobial is a sales feature in soaps, skin lotions, cleaning supplies, food preservatives, plastics, and even fabrics. However, only about one hundred species of microbes are known to actually cause diseases in humans; the vast majority of the thousands of species that inhabit us do not cause any problems, and, in fact, seem to come with serious benefits.

At first glance, our war on microbes, along with other medical advances, has truly paid off. In 1915 the average life span in the US was fifty-two years, about thirty years shorter than it is today. For better or for worse, there are almost four times more humans on this planet than there were just a hundred years ago, which translates to an incredibly accelerated growth in our historic timeline. Evolutionarily speaking, we've hit the jackpot. But at what price?

Revenge of the Microbes

The prevalence of infectious diseases declined sharply after the emergence of antibiotics, vaccines, and sterilization techniques. However, there has been an explosion in the prevalence of chronic noninfectious diseases and disorders in developed countries. One hears about these in the news all the time since they're very common in industrialized nations, where alterations to our immune system play an important role in their development. They include diabetes, allergies, asthma, inflammatory bowel diseases (IBDs), autoimmune diseases, autism, certain types of cancer, and even obesity. The incidence of some of these disorders is doubling every ten years, and they are starting to appear sooner in life, often in childhood. They are our new epidemics, our modern-day bubonic plague. (By contrast, these diseases have remained at much lower levels in developing countries, where infectious diseases and early childhood mortality are still the major problems.) Most of us know someone suffering from at least one of these chronic illnesses; due to this prevalence, researchers have focused their attention on identifying the factors that cause them. What we know now is that although all of these diseases have a genetic component to them, their increased pervasiveness cannot be explained by genetics alone. Our genes simply have not changed that much in just two generations—but our environment sure has.

About twenty-five years ago a short scientific article published by an epidemiologist from London attracted a lot of attention. Dr. David Strachan proposed that a lack of exposure to bacteria and parasites, specifically during childhood, may be the cause of the rapid increase in allergy cases, since it prevents proper development of the immune system. This concept was later termed the "hygiene hypothesis," and an increasing number of studies have explored whether the development of many diseases, not just allergies, can be explained by this hypothesis. There is now a large amount of very solid evidence, which we'll examine in the following chapters, supporting Dr. Strachan's proposal as generally correct. What remains less clear is what exact factors are responsible for this lack of microbial exposure. For his study on allergies, Dr. Strachan concluded that "declining family size, improvements in household amenities, and higher standards of personal cleanliness" contributed to this reduced contact with microbes. While this may be true, there are many other modern-life changes that have an even stronger impact on our exposure to microbes.

One of these changes can be attributed to the use, overuse, and abuse of antibiotics—chemicals that are designed to indiscriminately kill bacterial microbes. Definitely one of, if not *the* greatest discovery of the twentieth century, the emergence of antibiotics marked a watershed before-and-after moment in modern medicine. Prior to the advent of antibiotics, 90 percent of children would die if they contracted bacterial meningitis; now most cases fully recover, if treated early. Back then, a simple ear infection could spread to the brain, causing extensive damage or even death, and most modern surgeries would not even be possible to contemplate. The use of antibiotics, however, has become far too commonplace. Between the years 2000 and 2010 alone there was a 36 percent increase in the use of antibiotics worldwide, a phenomenon that appears to follow the economic growth trajectory in countries such as Russia, Brazil, India, and China. One troubling thing about these numbers is that the use of antibiotics peaks during influenza virus infections, even though they are not effective against viral infections (they are designed to kill bacteria, not viruses).

Antibiotics are also widely used as growth supplements in agriculture. Giving cattle, pigs, and other livestock low doses of antibiotics causes significant weight gain in the animals and, subsequently, an increase in the meat yield per animal. This practice is now banned in Europe, but is still legal in North America. It seems that antibiotic overuse in humans, especially in children, is inadvertently mimicking what occurs in farm animals: increased weight gain. A recent study of 65,000 children in the US showed that more than 70 percent of them had received antibiotics by age two, and that those children averaged eleven courses of antibiotics by age five. Disturbingly, children who received four or more courses of antibiotics in their first two years were at a 10 percent higher risk of becoming obese. In a separate study, epidemiologists from the Centers for Disease Control and Prevention (CDC) found that states in the US with higher rates of antibiotics use also have higher rates of obesity.

While these studies didn't prove that antibiotics directly cause obesity, the consistency in these correlations, as well as those observed in livestock, prompted scientists to have a closer look. What they found was astonishing. A simple transfer of intestinal bacteria from obese mice into sterile ("germ-free") mice made these mice obese, too! We've heard before that many factors lead to obesity: genetics, high-fat diets, high-carb diets, lack of exercise, etc. But bacteria—really? This raised skepticism among even the biggest fanatics in microbiology, those of us who tend to think that bacteria are the center of our world. However, these types of experiments have been repeated in several different ways and the evidence is very convincing: the presence and absence of certain bacteria early in life helps determine your weight later in life. Even more troubling is the additional research that shows that altering the bacterial communities that inhabit our bodies affects not just weight gain and obesity, but many other chronic diseases in which we previously had no clue that microbes might play a role.

Let's take asthma and allergies as an example. We are all witnesses to the rapid increase in the number of children suffering from these two related diseases. Just a generation ago it was rather unusual to see children with asthma inhalers in schools. Nowadays, 13 percent of Canadian children, 10 percent of US children, and 21 percent of Australian children suffer from asthma. Peanut allergies? That used to be incredibly rare, but is now so frequent and so serious that it has led to peanut-free schools and airplanes. As with the obesity research, it is now evident that receiving antibiotics during childhood is associated with an increased risk of asthma and allergies.

Our laboratory at the University of British Columbia became very interested in this concept and decided to do a simple experiment. As had been observed with humans, giving antibiotics to baby mice made them more susceptible to asthma, but what we observed next left us in awe. If the same antibiotics were given when the mice were weaned and no longer in the care of their mothers, there was no effect in susceptibility to asthma. There appeared to be a critical window of time, early in life, during which antibiotics had an effect on the development of asthma. When given orally, the antibiotic that we chose, vancomycin, kills only intestinal bacteria, and does not get absorbed into the blood, lungs, or other organs. This finding implied that the antibiotic-driven change in the intestinal bacteria caused the increase in the severity of asthma, a disease of the lungs! This experiment, as well as others from several different labs, came to the same conclusion: modifying the microbes that live within us at the beginning of our life can have drastic and detrimental health effects later in life. The discovery that this early period in life is so vulnerable and so important tells us that it's crucial to identify the environmental factors that are disturbing the microbial communities that inhabit us during childhood.

One of these factors has been observed by comparing children raised on rural farms to those raised in a city. Several studies have shown that exposure to a farming environment makes children less likely to develop asthma, even children from families with a history of asthma, and scientists are now beginning to learn why. Farm-raised children are exposed to more animals, more time outside, and a lot more dirt (and feces!), all things that are known to stimulate the immune system. A critical part of the training and development of the immune system occurs in the first years of life. Asthma, characterized by a hyperactive immune system, seems to have a higher chance of developing in a child with a limited exposure to these immune stimulants, because without them, the immune system does not have all the tools for proper development. By cleaning up our children's environments, we prevent their immune systems from maturing in the way they have for millions of years before us: with lots and lots of microbes. Life for our ancestors involved massive exposure to microbes from the environment, food, water, feces, and many other diverse sources. Compare that to our current way of life, where meat comes on sterile Styrofoam pans wrapped in plastic wrap, and our water is treated and processed until it's free of nearly all microbes.

Kids Will Be Kids

A friend, Julia, moved to a small free-range pig and poultry farm when her first child was a preschooler. She observed firsthand how differently a kid grows up in a city and on a farm. She has always been outdoorsy, so even when she was living in the city she would let Jedd, her oldest child, play outside a lot. They would go to parks and playgrounds, where she would encourage Jedd to get dirty, play in sandboxes and mud puddles—she even allowed him to put (safe-sized) objects in his mouth, like big rocks or leaves. Her outdoorsy nature, she thought, would make their transition to rural life easier, and it did in many ways. But nothing prepared her for the things she's seen her kids do on their farm. When her second baby was born, she would strap him on her back every morning so she could go to their chicken coop to pick up eggs. Jedd, timid with the animals at first, was now chasing and riding the chickens, tasting their feed and touching the fresh eggs. A couple of times she even caught him chewing on something he had picked up from the ground. Anyone who has stepped inside a chicken coop knows what's on the floor, so she's pretty sure Jedd has tasted chicken droppings at least a few times. Clearly, Julia freaked out at first, but it's hard to prevent a five-year-old boy from getting dirty when you're busy working and looking after a second child. After realizing that Jedd wasn't getting sick from his newly acquired tastes of the farm, Julia relaxed a bit. Jedd, now eight years old, is responsible for gathering the eggs every morning. Newly laid eggs are often soiled and he doesn't wear gloves. He washes his hands when he's done, but it's impossible that some of that stuff hasn't made it into his mouth.

Julia's second child, Jacob, was born and raised on the farm and, like his big brother, he was never the slightest bit hesitant to get dirty. He was once found playing knee-deep in a cesspool of pig waste. At fourteen months he swallowed a handful of fresh chicken droppings as Julia rushed towards him to prevent it. Her initial worry that her children were going to contract a disease from all this messiness dissipated as her kids remained healthy.

Nowadays, with her third baby strapped on her back, she doesn't even flinch at the sight of the two older boys doing what all farm kids do: getting very, very dirty. Every single day, they come home with dirt, poop, feathers, and who knows what else caked onto their skin and clothes. They try their best to keep their farm boots for outdoor use only, but it inevitably happens that dirty boots make it onto the living room carpet. Julia makes sure to wash their hands before they eat and they rarely miss a daily bath (the color of the bathwater is a constant reminder of why daily baths are mandatory in their house).

Even if they play outside a lot, most children growing up in urban environments rarely ever reach the level of dirtiness that Julia's kids experience on a daily basis. From this perspective, a farm kid (and his microbes) is very different from a city kid. We are by no means suggesting that we should all allow our kids to play with animal waste, as they could become sick from this. But farms in general provide a microbe-rich environment that has proven beneficial for the development of the immune system, and that really is akin to the way we used to live, which has been seriously altered only in the past few generations.

The vast majority of children have something in common with Jedd and Jacob, in that they all seek out dirt and enjoy getting messy and sucking on things. Why is that? Our natural behavior in the early years of life definitely tries to maximize our exposure to microbes: babies are in direct contact with maternal skin while breastfeeding, they are constantly putting their hands, feet, and every imaginable object in their mouths. Crawlers and early walkers have their hands all over the floor, and then in their mouths. It often seems that they're waiting for the few seconds that parents take their eyes off them to almost magically find and put the dirtiest thing they can reach in their drool-dripping mouths. It makes us wonder: Are kids instinctually drawn to microbes?

Older kids love digging in the dirt, picking up worms, rolling on the ground, catching frogs and snakes, etc. Perhaps this is actually natural behavior designed to populate kids with even more microbes. Children rarely hesitate to lick anything or anyone. As would be expected, children also suffer more infections than adults. Their vacuum-like behavior ensures that they taste the microbial world and subsequently train their immune system to react to it accordingly. If they encounter a disease-causing microbe, also known as a pathogen, their immune system detects it, reacts to it in the form of sickness, and then tries hard to remember it so that their body can prevent it from causing disease the next time this pathogen makes a visit. When the immune system encounters a harmless microbe—and the vast majority of microbes are harmless—it detects it and, through a series of mechanisms that science does not yet fully understand, decides to ignore or tolerate it. Thus, if children's lifestyles and behaviors dictate a limited exposure to these training events, their immune system will be partially immature and will not learn how to properly react to a pathogen or how to tolerate harmless microbes. The consequence of missing out on this early training appears to be that, later in life, the immune system may react too fiercely to these harmless microbes, which could trigger inflammatory responses in various organs of the body. This contributes to the appearance of "developed country diseases" (like asthma and obesity) that are becoming so prevalent today.

Microbes to the Rescue

Helping develop our immune systems is only part of what microbes do for us. They are in charge of digesting most of our food, including fiber and complex proteins, and chopping them into more digestible forms. They also supply the essential vitamins B and K by synthetizing them from scratch, something our own metabolism cannot do. Without the vitamin K from microbes, for example, our blood would not coagulate.

Good bacteria and other beneficial microbes also help us combat disease-causing microbes. Experiments in our lab have shown that infections from *Salmonella*, a diarrhea-causing bacterium, are far worse when antibiotics are given before the infection actually occurs. Similarly, many of us have experienced the side effects of a long bout of antibiotics: abdominal cramps and watery diarrhea. The microbes we harbor live in a balanced state that provides us with so many benefits, all in exchange for a portion of our daily calories and a warm, dark place to live with regular feeding and watering.

But changes in our modern lifestyles are altering this balance, especially during a critical window in early life. In many developed countries, about 30 percent of babies are born by cesarean section, antibiotic usage is a lot more frequent, and most children do not suffer serious infections thanks to vaccines. Far from suggesting that any of these things should be avoided, our aim is to educate parents, as well as parents-to-be, grandparents, and caregivers, about the potentially life-changing decisions we make on a daily basis by raising children in an environment that's much cleaner than ever before. As parents ourselves, we understand that most of us do the best we can with what we have, and it is not our intention to dictate how other people should raise their children. However, as microbiologists, we are becoming increasingly aware of the key role our resident microbes have in shaping our bodies' development. The microbial communities of babies and young children are being altered in ways that may make them sicker later in life, by the very same practices intended to keep them healthy. Talk about a double-edged sword!

The scientific community is just beginning to grasp this new knowledge, and the general public is just starting to hear about it in news articles of (often misinterpreted) studies. Preventing serious illnesses should always be one of our biggest concerns, but we can also do a great deal to try to distinguish between a necessary intervention, such as giving an antibiotic to fight a life-threatening bacterial infection, and an unnecessary and hyperhygienic practice, such as applying antimicrobial hand sanitizers every time a child plays outside. Not all children will or should be raised like Jedd or Jacob, but we can certainly change those unneeded aspects of our far-too-clean world.

In our classical training as microbiologists, we studied only the microbes that cause diseases and the ways to kill them. Now we acknowledge that we have, for many years, ignored the vast majority of microbes that keep us healthy. Our research labs are changing focus, and we are beginning to think it's time for everyone to become better hosts to our microbial guests.

BUBBLE BOY

David Vetter was born in 1971 in Houston, Texas, with a rare genetic disorder that left him without a working immune system. Any contact with a nonsterile world would mean certain death. Because of this, he was delivered by C-section and placed in a sterile bubble immediately after his birth. In a controversial medical decision, he lived in the hospital in a bubble that grew with him. His medical treatment included many courses of antibiotics to prevent any bacterial infection. Being devoid of bacteria meant that doctors also had to feed him a special diet, along with the essential vitamins K and B, which are normally produced by intestinal bacteria. David's story reflects the impossibility of living without an immune system in a world full of microbes, as well as a human's dependence on microbes and what they produce for us. Sadly, David died at the age of twelve from a viral infection a few months after a bone marrow transplant was finally performed.

2: A Newly Discovered Organ: The Human Microbiome

Invisible Life

The idea of humans being inhabited by countless microbes invisible to the naked eye is as old as the first microscope. Born in 1632 in the city of Delft, in what is now the Netherlands, Antoni van Leeuwenhoek was a tradesman with a special interest in lens making. His desire to see the intricacies of the cloths he marketed drove him to shape glass rods into spheres using a flame. These almost perfect spheres allowed him to magnify not just threads, but anything else he wanted to view in great detail. Although he wasn't formally trained as a scientist, he was one at heart and he soon began to put the oddest things under his rudimentary microscopes: water from a creek, blood, meat, coffee beans, sperm, etc. He methodically wrote everything down and sent his findings to the Royal Society of London, which began publishing his curiosities-filled letters.

One day in 1683, he decided to scrape the white residue between his teeth and place it under his lens, writing in his notes:

An unbelievably great company of living animalcules, a-swimming more nimbly than any I had ever seen up to this time. The biggest sort (whereof there were a great plenty) bent their body into curves in going forwards . . . Moreover, the other animalcules were in such enormous numbers, that all the water . . . seemed to be alive . . . All the people living in our United Netherlands are not as many as the living animals that I carry in my mouth this very day.

Naturally, Leeuwenhoek's observations of a never-before described world filled with microscopic "animalcules" were met with great skepticism and ridicule. It wasn't until other British scientists saw it with their own eyes that they began to acknowledge that Leeuwenhoek was not hallucinating. Leeuwenhoek had written many letters to the Society, but discovering microscopic life is what sealed his long-lasting fame. As a result of his many discoveries, Leeuwenhoek is considered the "Father of Microbiology."

Still, these findings remained nothing more than curiosities of the natural world, with no real connection to human biology until scientists discovered that those "animalcules" caused diseases. This revelation took place almost two hundred years later, when Robert Koch, Ferdinand Cohn, and Louis Pasteur each separately confirmed that diseases such as rabies and anthrax were caused by microbes. Pasteur's work also showed that microbes caused the spoilage of milk, and he thus designed the process known as pasteurization, in which microbes are killed with the use of high heat. Milk contamination led Pasteur to the idea that microbes could be prevented from entering the human body, and together with Joseph Lister, they developed the first antiseptic methods. These began to be widely adopted, with one of them still in use today: Listerine.

Avoiding Contagion at Any Cost

The work of Pasteur, Cohn, Koch, and others led to the widespread knowledge that diseases could be avoided by preventing contact with microbes, and by killing them, and so the quest to eradicate them began in earnest. Health departments opened in London, Paris, New York, and other big cities. Garbage, which had previously been left to pile high on sidewalks, was now collected and disposed of; drinking water was treated; rats and mice were hunted; sewer systems were built; and people with contagious diseases were often placed in isolation. It was through all this that the word "bacteria" gained its bad reputation and inherent connotation of disease, contagion, and plague. Germs were (and still are) entities to be feared, avoided, and fought.

Fast-forward another two hundred years and an equally astounding discovery is now in progress: in our quest to clean up our world, we have been killing more microbes than necessary and, ironically, this can make us sick. Why? Because our bodies know how to properly develop only in the presence of lots of microbes. This groundbreaking concept significantly expands on what science already knows about the nonharmful bacteria that inhabit our body: that they aid in the digestion of certain foods, and that they fabricate certain essential vitamins. However, only very recently have we begun to comprehend how profoundly necessary microbes are for our normal development and well-being.

Microbes: Partners in Evolution

The last twenty years of studying microbes has allowed us to understand that microbes aren't optional forms of life that live within us; they truly constitute part of who we are biologically. To get a better grasp on this, we must first understand that our partnership with microbes is as old as the first species of hominids (our ancestors), and that the evolutionary changes that hominids experienced were accompanied by changes in our microbiota, too. Throughout human history there have been only a few landmark evolutionary bursts (rapid evolutionary changes) that have marked the course of hominids. Interestingly, two of them can be clearly linked to changes in our intestinal physiology and thus with our microbiome.

As hunters and gatherers (a lifestyle that lasted about 2.5 million years), our ancestors had no permanent homes, living in temporary shelters with few possessions so they could easily move from one place to another. Depending on the geographic region they inhabited, early humans ate different mixtures of meats, roots, tubers, and fruits—whatever was in season. Then an extremely important event occurred that led to one of these evolutionary bursts: our ability to control fire and cook food. We completely take it for granted now, but cooking food made it safer to eat, as heat kills the disease-causing bacteria that thrive in decomposing meat. It also changes the chemistry of the food itself, making it much easier to digest and a lot richer in energy. This sudden increase in energy levels changed everything for humans. No longer did our ancestors have to spend hours chewing raw food in order to extract enough calories to sustain everyday life. Think of what our closest relatives in nature, apes, are almost always doing when see them in the zoo or on TV. If humans hadn't developed a way to cook food we, too, would have to spend six hours chewing five kilos of raw food every day to get enough energy to survive, just like our primate cousins do.

The fossil records of humans from this period consist of bones and teeth, making it impossible to determine what type of microbiota lived in the intestines of ancient hunters and gatherers. However, anthropologists have been able to show that the change in lifestyle and diet that resulted from the advent of cooking had anatomical consequences involving the intestines. As energy intake increased, the intestines of our human ancestors shortened and, amazingly, their brains grew, too, increasing in size by about 20 percent. Given what we know today about the link between gut microbes and brain development, it is very likely that intestinal microbiota had a part in this "sudden" brain growth. Brain enlargement improved our capacity to hunt, communicate, and socialize. In other words, cooking made us smarter—it made us human.

Another evolutionary landmark occurred about eleven thousand years ago. Certain groups of humans realized, probably by chance, that fallen grains from the wild wheat stalks they collected would give rise to more wheat if planted. When humans learned to domesticate plants for food, they tossed away their nomadic ways for a settled lifestyle. Having crops nearby meant that previously small tribes of a few dozen humans could grow to a few hundred, which in turn gave rise to basic traits of civilization, such as trade, written language, and math. If it weren't for farming, we would all still be picking berry after berry from bushes and walking miles every day. The emergence of agriculture coincides with the appearance of the first cities; inadvertently, agriculture built our modern social structures. This lifestyle change was so successful that farmers replaced foragers, and these days only a handful of people maintain a hunter-gatherer way of life.

As expected, the lifestyle associated with farming came with major dietary changes. Humans no longer ate small bites throughout the day with the occasional feast after a hunt since farmers had a steady and somewhat predictable supply of foods. So how did this affect our microbiota? By domesticating grains and consequently obtaining most of their daily calories from their new crops, the diet of farmers became less diverse. Based on what is currently known about the microbiota's response to diet, their microbiota likely became less diverse, too. In fact, comparing the intestinal microbiota of the Hazda people of Tanzania, one of the few contemporary tribes that relies on foraging, to a modern farmer is like comparing a rain forest to a desert, in terms of biodiversity. Less diversity in our microbiota is associated with a number of human diseases, many of which we cover in later chapters.

Although farming has been around for only eleven thousand years (just 0.004 percent of human history!), physiological changes have also been linked to the agricultural diet, and some of these changes involve our resident microbes. The new diet brought with it cavities and other periodontal diseases, mediated by bacteria rarely found in foragers. Our teeth, jaws, and faces have grown smaller, too, probably because chewing was reduced on such a diet. Some evolutionary biologists believe that we lived a healthier lifestyle as foragers, and that humans traded in that healthier lifestyle for food security and more babies (not a bad deal, actually!). Certain nutritionists have extrapolated from this a recommendation that, in order to promote health, all modern humans should eat the way hunters and gatherers did, but this has been debunked by top evolutionary biologists based on the fact that humans have adapted genetically to the challenges that farming generated (see the Caveman Diet, page 30).

What these two major events in human history teach us is that changes in lifestyle are accompanied by changes in our microbiota, and that these microbial changes might affect our health for better (e.g., cooking food and decreasing infections) or worse (e.g., agriculture and less microbial diversity). Whether we like it or not, we are married to microbes for life, in sickness and in health, for richer or for poorer.

Bugs "R" Us

Our microbes are part of what make us human, but our current way of living and eating, especially in the Western world, has exerted further changes in our microbiota and in our biology. In the past hundred years, and especially the last thirty years, humans have learned to process foods to make them tastier, more digestible, and more shelf-stable than ever before. On top of this, our push to clean up our world in order to fight infectious diseases, including the use of antibiotics, has further shifted the composition and diversity of our microbial communities. Double-punching our microbiota like this has induced huge changes in our intestinal environments and, as we will learn in the following chapters, on many other aspects of our bodies' normal functions.

In order to appreciate how the microbiota influences our health, it is important that we discuss certain basic biological concepts about our microbiota and the organ most of them call home, the human intestine. The human microbiota consists of bacteria, viruses, fungi, protozoa, and other forms of microscopic life. They inhabit our skin, oral and nasal cavities, eyes, lungs, urinary tract, and gastrointestinal tract—pretty much any surface that has exposure to the outside world. Another term that is frequently used is microbiome, which refers not only to the identity of all the microbes living within us, but also to what they do. A total of 10^{14} microbes are estimated to live in the human body and, as mentioned, the intestinal tract is the biggest reservoir of microbes, harboring approximately 10^{13} bacteria. It is this community that influences us, their host, the most. In fact, unless otherwise noted throughout this book, when we use the term microbiota, we are referring to the intestinal microbiota. Although bacteria are approximately twenty-five times smaller than human cells, they account for a significant amount of our weight. If we were to get rid of our microbiota we would lose around three pounds, or about the weight of our liver or brain! A single bowel movement is 60 percent bacteria numbering more than all the people on this globe, a deeply disturbing fact for germophobes.

For microbes, the gastrointestinal system is a fabulous place to live. It's moist, full of nutrients, and sticky (allowing microbes to adhere to it), and in many sections it completely lacks oxygen. Although it seems counterintuitive that any life-form would favor a place without oxygen, an enormous number of bacterial species either prefer or require such a place, as this world evolved for billions of years without oxygen. Microbes living without air are called anaerobes and our gut is packed with them.

About 500–1,500 species of bacteria live in the human gut; the types and numbers vary according to the different sections of the gastrointestinal system. Starting from the top down, the mouth harbors a diverse and complex microbiota—the tongue, cheeks, palate, and teeth are all covered in a dense layer of bacteria known as a biofilm. For example, the dental plaque that dentists remove from our mouths is one of these biofilms. The stomach, on the other hand, is not the best place for microbes, as it is as acidic as battery acid. Still, a few bacterial species have adapted to live under such conditions. Farther down are the small and large intestines, where the number of microbes continues to increase until we reach the very end of the large intestine. Oxygen follows the opposite pattern, as it gradually decreases towards the lower portions of the gut, allowing strict anaerobes (those that die when exposed to the slightest bit of oxygen) to flourish in the large intestine. The differences in living conditions within the small and large intestines determine the number and the types of bacteria that reside in each portion of the gut. For example, the slightly acidic and oxygenated environment in the upper small intestine allows for bacteria that are tolerant to these conditions, such as the bacteria we often eat in our yogurt, known as *Lactobacilli*. Unlike the upper small intestine, the large intestine, also known as the colon, moves or churns its contents very slowly and produces a lot of mucus, allowing for many more bacteria to grow, especially those that use mucus for food.

Another characteristic of the human microbiota is its variability between individuals. Although about one-third of bacterial species are shared between all humans, the rest of them are more specific, making our microbiome unique like a fingerprint. Similarities in microbiota are highly dependent on diet and lifestyle, and to a lesser extent, on our genes. For example, identical twins (who share all of their genes) can have very different microbiotas if one is a vegetarian and the other eats meat. Family members, including husbands and wives who are not genetically related, tend to have similar microbiotas due to a shared living environment and diet. Humans also have striking similarities with the microbiotas of several species of apes, but only those that are omnivores like us. Mountain gorillas, for example, have a microbiota much more closely related to pandas, because they both spend their days leisurely eating bamboo.

Once established in our intestine, microbial communities are very stable. Only drastic changes, such as adopting a vegan lifestyle or moving to a completely different part of the world, will significantly alter your microbiota. Going on antibiotics for a week to treat an infection will also affect your microbiota, but only temporarily in most cases. It will generally bounce back to something resembling its pre-antibiotic state after you finish the treatment and go about your old way of eating. However—and this is a big however—the microbiota takes about 3–5 years from the time we're born to become a fully established community, and during this period it's very unstable, especially during the first few months of life. Any drastic changes to it have a very high chance of altering the microbiota permanently. In fact, it is the early colonizers of the intestinal microbiota that have a major influence on the type of microbiome we have later in life. Thus, a short-lived event like a C-section may have long-lasting consequences, since a baby born this way starts with a very different microbiota than a baby born vaginally. The potential health outcomes and impact of this type of event during early life has major implications for later health and disease, as discussed in later chapters.

Immune Cell School

Given the strong associations between early-life alterations to the microbiota and immune diseases later in life, we might ask: What exactly are microbes doing to us when we're babies that is so important? As mentioned in the previous chapter, microbes help us use food that we can't digest properly, and they also fight off bacteria capable of causing us harm. We've known about these roles for decades, but they are just the tip of the iceberg. As soon as we're born and begin getting colonized with bacteria, bacteria kick-start a series of fundamental biological processes in our body. One of them is the maturation of the immune system, the network of cells and organs that defend us from diseases.

Before scientists started unraveling the role of the microbiota in immunity, every doctor and scientist was taught that we're born with an immature immune system that gets trained in a small organ called the thymus. Here, immune cells known as T cells—the strategists of our immune system—are taught who is a friend and who is a foe. This training boot camp lasts for a few years only, until the thymus disappears, and all our immune cells have acquired this knowledge. Immunologists deciphered a complex series of mechanisms showing exactly how this occurs, but they couldn't explain one big question: How does the thymus teach immune cells which kinds of bacteria are beneficial and which ones aren't? After all, since we're covered head to toe (also inside and out) with microbes, mostly good ones, how do immune cells know the difference? The thymus does not interact with bacteria, so where could it get this information? It turns out this very important aspect of the training doesn't occur in the thymus—it happens in our gut.

Before we're born, the lining of our gut is full of immature immune cells, and as soon as we come into the world and bacteria start moving into their new home, these immune cells "wake up" almost magically. They start multiplying, they change the type of activities they do, and they even move to other parts of the body to train other cells with the information they just received. Experiments with germ-free mice, which are mice that are born into and kept in a completely microbe-free environment, show that without microbes the immune system remains immature, sloppy, and unable to fight off diseases properly.

Scientists haven't figured out exactly how microbes do this at the molecular level, but it is known that most bacteria will teach these immune cells to tolerate them, whereas some bacteria—the pathogens that cause disease—have the opposite effect. This makes sense; if our immune cells started fighting off all bacteria indiscriminately, there would be an out-of-proportion inflammatory battle between the small quantity of immune cells and the vast numbers of bacteria right after we're born. In reality it's quite the opposite; despite the enormous amount of bacteria living in the intestine, it's a relatively controlled and harmonious place. The way this is achieved is by the microbiota modulating the immune system, allowing most microbes to be tolerated.

Many inflammatory diseases, such as asthma, allergies, and IBD, are characterized by an overreactive immune response. Knowing what we do now about the importance of microbiota in immune system development, it's not surprising that these diseases are being diagnosed in more and more children. They are, to a great extent, a consequence of the modern lifestyle changes that are altering the types of microbes that affect the immune system. There's a reason immune cells wait for microbes to come and train them right after we're born: because this is the way it has happened for millions of years and is the way it will always be. We need to find ways to modify our modern behavior so that immune cell school can function properly.

Feeding Our Microbes So
They Can Feed Us

Another fundamental function of microbes is to aid in the regulation of our metabolism. Humans, just like any other living animal, obtain energy from food that is digested and absorbed in the intestines. Besides helping us digest certain foods that the intestines can't handle on their own, bacteria produce energy for us, and the amount they produce is noteworthy. Germ-free mice weigh significantly less than conventionally raised mice, but once bacteria begin to colonize them they have a 60 percent weight gain, despite not eating more food than regular mice. One of the mechanisms by which they accomplish this is a process known as fermentation. Think of the intestine as a bioreactor where bacteria ferment fiber, carbohydrates, and proteins that were not digested and absorbed in the small intestine. The end-products of this process are called short-chain fatty acids (SCFA), and three of them are very important to different aspects of human energy metabolism: acetate, butyrate, and propionate. Intestinal cells rapidly absorb SCFA and use them as an energy source to stay fueled. SCFA are also transported very rapidly to the liver, where they are transformed into critical compounds involved in energy expenditure and energy storage. SCFA help determine how and when we use the energy obtained from food, and, importantly, when to store it as fat. Thus, it's not surprising that alterations in the production of SCFA have been associated with obesity, both in mice and in humans.

SCFA are not exclusively produced by the microbiota. These compounds are too critical for our metabolism to rely entirely on bacteria for their production. Still, studies performed on patients genetically unable to produce propionate have shown that approximately 25 percent of the propionate in our body is derived from bacterial activity in the gut. The implications of this are significant, considering that treatment with many types of antibiotics severely alters intestinal SCFA production. If antibiotics are given during early childhood, especially in the first few months of life, the risk of experiencing long-lasting metabolic and immune alterations due to abrupt changes to the microbiota increases dramatically.

Scientists haven't yet figured out all the functions that our metabolism delegates to the microbiota. Immune training and metabolizing energy are two essential things that our microbes do for us, but it's clear that there are more. Brand-new research shows that the microbiota plays an important role in neurological development (discussed in chapter 15), and even in the health of our blood vessels. These types of discoveries have led scientists to call our microbiome a "new organ," perhaps the last human organ to be discovered by modern medicine. Although most of this knowledge has just recently emerged and many pieces of the puzzle remain unsolved, it is evident that protecting the initial developmental stages of our microbiota has a significant impact in human health.

In the next four chapters we discuss the life stages that are most influential in the development of the human microbiome, all of which occur during infancy and early childhood. We will explore how some of the actions parents take during pregnancy and birth, as well as through diet, can have profound implications in the communities of microbes that are part of our children's bodies. With scientific information parents have learned to make better choices when raising their kids, such as limiting sugar intake and even the amount of time spent in front of the TV. With our newfound awareness of how important the microbiome is, let's explore what we might do as parents to improve our children's health by caring for their microbes.

THE CAVEMAN DIET

The newest diet fad suggests that eating the way our Paleolithic ancestors did will make us be healthier and live longer. However, evolutionary biologists don't agree with this because it's not based on current scientific knowledge. Some assumptions of the "paleo diet" include:

- *Our ancestors ate mostly meat, and no legumes or grains.* Actually, our ancestors ate incredibly different diets depending on where they lived. One could expect this statement to be close to the truth in Arctic environments, but in more temperate weather this was not the case. Biochemical analysis of dental fossil records from this period show that foragers did eat grains and legumes. Also, the meat we consume today—from domesticated livestock—is completely different than the wild game our ancestors ate.

- *Our ancestors did not eat dairy.* While this is generally correct, modern humans from many regions of the world where dairy is consumed have genetically modified their metabolism to digest and absorb dairy products. In other words, we have evolved, in a somewhat short period of time, to digest foods that our ancestors didn't eat. Our genes have changed since we roamed the savannahs.

It is impossible for modern humans to eat the way our ancestors did because our foods today are completely different than before. Carrots, broccoli, and cauliflower did not exist back then, and neither did the leaves used to make salads. All of these are products of agriculture. What certainly is true is that the typical modern human diet has extremely low diversity and is heavily processed, compared to food consumed a hundred years ago.

In addition, only very recently have people stopped eating just what is in season and whole foods. These are the dietary changes that really have an impact on our health, in great part because of the effects on our microbiota. Yes, eating fewer refined carbohydrates and more vegetables will help you lose weight and feel better, but this does not reflect our Paleolithic past in the way "paleo" enthusiasts believe it does.

PART TWO

Raising Babies and Their Microbes

3: Pregnancy: Eating for Two? Try Eating for Trillions

The Pregnant Microbiota: Another Reason to Eat Well

Seeing that positive result on a pregnancy test changes everything for most women. All of a sudden they're going to the bathroom more times than they can count, forgetting where their keys are while they're holding them in their hands, falling asleep at work (at 10 a.m.!), feeling full right after a meal, only to feel famished ten minutes later. From differences in her skin and hair to buying pants in three sizes within one year, pregnancy is a time of major changes in a woman's body. In nine short months, a woman undergoes a series of drastic physiological transformations that nurture a single fertilized cell into a crying, hungry baby. Many of our organs alter their functions to facilitate these new biological needs of both the mother and her developing baby. For example, the liver produces 25–35 percent more fats in order to promote baby growth. Fats, also known as lipids, are formed as a way to store energy. By naturally adjusting liver metabolism to make more lipids, a pregnant mother's body ensures that there will be enough energy for the baby to grow, and for the future production of milk following delivery.

Like the liver, a pregnant woman's microbiota also responds to this new state. In fact, experts believe this change is a normal physiological adaptation to support the growth of the fetus. A recent study showed that the microbiota of a pregnant woman in her third trimester strikingly resembles the microbiota of an obese person (just what every pregnant women wants to hear . . .). Moreover, when the microbiota of a female mouse in late pregnancy was transferred into a germ-free mouse, the latter mouse gained a lot of weight, despite not increasing food intake or being pregnant. This study was carried out in the laboratory of Dr. Ruth Ley at Cornell University in New York, a scientist at the forefront of the microbiota field. She believes that late pregnancy is an energy-thirsty period, during which the body takes advantage of the energy-producing machinery of the microbiome to promote weight gain for the benefit of the mother and her baby. The timing for this large shift in microbiota couldn't be better, occurring towards the end of the pregnancy when babies start packing on the pounds and when women need to start preparing for the energy demands of breastfeeding.

This same study, which sampled ninety-one pregnant women (the largest to date), also showed that some species of bacteria that were more predominant in the third trimester of pregnancy were also found in their babies at one month of age. This suggests that another consequence of the big change in microbiota during pregnancy is to pass many of these bacteria on to the newborn. It's fascinating to think that a woman's body and her microbiota work together during pregnancy, likely because both benefit from having a new baby. From a genetic perspective, having babies is the only way to propagate our genes; from a microbial perspective, a newborn is brand-new real estate where microbial genes can also multiply and propagate.

Another recent study showed that the shifts to microbiota during pregnancy reflect the amount of weight women gain. According to the American Institute of Medicine, a woman of normal weight should gain 25–35 pounds during pregnancy, underweight women should gain 28–40 pounds, and overweight women should gain only 15–25 pounds. Women who gain more weight than what is considered standard have distinct changes in their microbiota. Given that a baby inherits many of its mother's microbes, and that some of these microbes actually promote weight gain, should we worry about passing obesity-associated microbes to our babies? Unfortunately, yes. Women need to watch their weight during pregnancy, especially during the last trimester. Obesity is a complex condition arising from both genetic and environmental (including microbial) factors (discussed in chapter 10), but it appears that even in cases in which obesity is considered genetic, microbes have a role in its development. This makes sense, as microbes are directly involved in the way we break down food and store fats. If you think no one is watching when you give in to that midnight snack craving, that's sadly not the case—microbes are watching what we eat at all times, since it affects them directly!

The good news is that, just as we can foster weight-gain microbes through a poor diet, we can promote the growth of beneficial microbes through a healthy diet. Although scientists haven't identified specific microbes associated with leanness yet, it has been shown that a varied diet that includes fruits, vegetables, and fiber promotes a diverse microbiota, a characteristic of lean (and healthy) individuals. Thus, you, and your microbiota, are what you eat—and there is probably no better time to watch your diet than when you're pregnant. Bad dietary choices during this stage of life will not only make women gain more weight than what is considered healthy, they also have the potential to influence a child's future ability to control weight. So, next time you walk by a candy machine, don't listen to your sugar-loving microbes, and nourish the trillions of microbes that are begging you to grab a piece of fruit instead.

The Vaginal Microbiota

During pregnancy, microbiota adaptation also occurs in the vagina, an organ that hosts millions of microbes. The composition of this microbiota influences vaginal health tremendously. Many women develop yeast infections after being on antibiotics or oral contraceptives (birth control pills alter the pH of the vagina). Bacterial vaginal infections, also known as vaginoses, are very common. These infections occur when yeast (often *Candida*) or bacteria overrun a beneficial group of microbes known as *Lactobacilli*, a type of lactic acid bacteria that is very common in the vagina. Lactic acid bacteria are also used in the dairy industry for the production of yogurt, kefir, cheese, and buttermilk. Many of them have health benefits and are used as probiotics.

During pregnancy, the number of vaginal *Lactobacillus* increases dramatically, which is thought to occur for two important reasons. First, by keeping the vagina acidic, the presence of *Lactobacillus* helps discourage disease-causing microbes such as *E. coli*, which do not like to grow in acidic conditions. There's probably no better time to arm the bacterial vaginal defenses than during pregnancy, when a pathogen could track up from the vagina, through the cervix, and into the uterus, where the baby is growing. In fact, it is known that certain vaginal infections during pregnancy are associated with pre-term and low-weight births. Second, *Lactobacilli* are great at digesting milk, as their name suggests (*lacto* is Latin for "of milk," and *bacillus* is the name given to rod-shaped bacteria). By ramping up the levels of *Lactobacillus* in vaginal secretions, more of these bacteria will reach the baby's gut (when born vaginally), and facilitate the digestion of the only food the baby will eat for months: her mother's milk. In this sense, *Lactobacilli* are probably a baby's first and best microbial friend.

The vaginal microbiota plays a very important role during pregnancy and birth, as it is one of the sources (along with the gut microbiota) of the first microbes to set up camp in a newborn. As soon as a baby is born vaginally, she gets covered in vaginal secretions and, yes, with fecal matter, too. Consequently, the composition of vaginal secretions is of utmost importance during pregnancy, and vaginal health should be taken very seriously during this period of time. Just as women should take care of their diet to promote a healthy intestinal microbiota, they should look after their vaginal health, too.

To promote vaginal health, gynecologists recommend that pregnant women wear cotton underwear, avoid vaginal douching (never recommended), avoid vaginal cleaning products, and use gentle, unscented soaps to clean the outside of the vagina only. The vagina is an organ that cleans itself through the production of secretions and needs little extra hygiene. In fact, cleaning the interior of the vagina is strongly associated with infections, as it alters the balance of the resident microbiota. In addition, it has been shown that the consumption of probiotics containing *Lactobacillus acidophilus* decreases vaginal infections. Several clinical studies suggest that eating yogurt may help, too, although not to the same extent as probiotics alone. You can even get probiotic preparations in the form of vaginal suppositories, which are used to treat such infections. Safe sex is the best way to avoid sexually transmitted infections (STIs); it is a practice that should always be followed, and especially during pregnancy. An STI contracted during pregnancy can be more dangerous to the mother than an STI contracted at another time, as immune systems are weaker during pregnancy—a physiological adaptation meant to prevent a woman's immune system from reacting to the fetus. Unfortunately, this makes a mother-to-be more vulnerable to infection.

Stress, Your Baby, and Your Microbes

Another important measure to maintain a balanced microbiota during pregnancy is to avoid stress, which is always easier said than done. We've all felt it—stress is a condition that affects most people at some point or another. It can be helpful sometimes, like when it compels you to finish an assignment for work that's due the next day. The problems arise when stress becomes an everyday companion; this is when it affects our health. Stress can make you lose sleep, have headaches or stomachaches, overeat, or lose your appetite. While pregnancy is typically a very joyful time, it can also be difficult. Dealing with the physical discomforts such as nausea, exhaustion, and backaches may quickly add up. On top of that, hormonal changes affect mood and the ability to handle stress.

A moderate level of stress is unlikely to cause a major impact on the health of a mother or her baby. However, certain situations may lead to severe stress, which can have detrimental effects on the pregnancy and the health of the baby. Abrupt negative life events, such as divorce, serious illness, financial problems, partner abuse, depression, and the conflicting feelings surrounding an unplanned pregnancy—to name a handful—are all causes of long-lasting or severe forms of stress. Some women suffer severe stress and anxiety when faced with the idea of labor or parenting. Severe stress is associated with preterm and low-weight births, and with certain illnesses in children, including skin conditions, allergies, asthma, anxiety, and even attention-deficit hyperactivity disorder (ADHD; see chapter 14).

A recent study from the Behavioural Science Institute of Radboud University, in the Netherlands, suggests that the microbiota plays a leading role in the link between stress during pregnancy and the aforementioned disorders. This study, which recruited fifty-six pregnant mothers, found that women who experienced high and prolonged levels of stress had alterations in the vaginal microbiota that could also be detected in their babies' gut microbiota. Infants born to highly stressed mothers showed lower levels of beneficial microbes, such as lactic acid bacteria. In the same study, these changes to the microbiota were associated with more gastrointestinal issues and allergic reactions in babies. They also found that the negative effects of severe maternal stress could not be corrected by breastfeeding, even though it has been repeatedly shown to promote a healthy microbiome in infants.

A similar study aimed at exploring the link between maternal stress and the microbiota was recently performed in mice. The study showed that a reduction in vaginal lactic acid bacteria, caused by stress, is accompanied by decreased immune functions in the offspring. Furthermore, the changes in the baby mice were not limited to the types of bacteria growing in their guts; there were also important metabolic differences detected in their blood and their developing brains. It may well be that the vaginal microbiota is at the center of this, responding to maternal stress and transferring its imbalanced state to the newborn, where it can lead to lasting health consequences. Although a casual relationship remains to be established, it appears that lactic acid bacteria from vaginal secretions are not only involved in facilitating milk digestion in newborns, but also carry out important metabolic functions in the developing newborn—yet another reason to reduce stress as much as possible and to take daily probiotics during pregnancy.

Infections and Antibiotics: Can We Avoid Them?

Controlling your diet and your stress levels during pregnancy is an enormously challenging goal for most women, but it can be done. However, the microbiota of pregnant women can suffer a big blow through a situation that's out of their control: taking antibiotics to treat an infection. As mentioned before, pregnant women are more vulnerable to infections, and if they occur, they are likely to be more severe due to their compromised immune systems. This is why it's recommended that pregnant women wash their hands often, avoid caring for people with infections (good luck with that when you have other kids!), avoid gardening without gloves, cook meats thoroughly, avoid changing the cat litter box, and avoid deli meats, sushi, and unpasteurized milk. Pregnancy is definitely not the time to get dirty and eat dirt, as we will later suggest our kids should do (although some pregnant women have an urge to do so—see Care for a Spoonful of Soil? on page 51).

Despite best efforts to avoid them, infections during pregnancy are quite common, with urinary tract infections (UTIs) and bacterial vaginoses both affecting about 1 in 6 pregnant women in the United

States and about 1 in 10 pregnant women in Canada. Other commonly diagnosed infections during pregnancy are respiratory tract and skin infections. Fortunately, several antibiotic medications are safe to use during pregnancy, but they're being prescribed to a lot of women—very likely more than necessary. The most recent National Birth Defects Prevention Study in the US, which has been collecting data since 1997, showed that almost 30 percent of women receive at least one course of antibiotics during pregnancy. A population-based study (a term given to studies involving a very large number of people) in the UK showed that the same is true for British women, while 42 percent of French and 27 percent of German women take antibiotics while pregnant. There's no debate about the immense change that an antibiotic brings to the microbiota. After a course of antibiotics, the overall diversity of the microbiota is substantially reduced. Its effect can be compared to what happens when a lush rain forest gets chopped down, and only a few dominant species make a comeback. Fortunately, the adult microbiota is fairly stable, and after finishing a course of antibiotics, in a nonpregnant woman this microbial forest usually returns to normal. The concern during pregnancy is that the microbiota fluctuates considerably, which is a characteristic of unstable ecosystems that are more susceptible to abrupt changes and permanent damage. When expectant women take antibiotics, especially in the last two trimesters, their microbiota takes a major hit, and according to new research, so does the microbiota of their babies. What becomes even more concerning is that antibiotic use during pregnancy is now being associated with certain diseases seen later in children.

A study of more than seven hundred pregnant women from New York showed that children born to those who received antibiotics in their second and third trimesters had an 85 percent higher risk of childhood obesity by age seven. These results are very significant because they were obtained after correcting for other confounding variables of obesity, such as the weight of the mother, the birth weight of the child, and whether or not the infant was breastfed. All of these factors were previously shown to be associated with the risk of obesity, so it's important (for this and any other similar study) to remove these variables from the analysis. These findings are quite new (published in 2014) and they still need to be replicated, but if more studies show a similar trend, it suggests that childhood obesity may have roots in the very early stages of human development, and that antibiotic use during pregnancy has significantly more risk than is currently assumed in medical practice.

Antibiotic use during pregnancy has also been associated with asthma, eczema, and hay fever in infants. Two large studies from Finland, a country that has experienced a twelve-fold increase in asthma rates since the 1960s, showed that using antibiotics during pregnancy is a significant risk factor for early asthma in babies. Other epidemiological studies have found similar associations between antibiotic use during pregnancy and inflammatory bowel disease (IBD) and/or diabetes, each of which is discussed in detail in forthcoming chapters. What's very peculiar is that these diseases share common risk factors. They are all immune disorders that have become increasingly common in the past few decades, and they usually occur in individuals with certain known genetic predispositions. Recent research on humans and animals show that the risk factors associated with these diseases also involve the early microbiota. How early? According to the studies, these changes begin before we're born, through mechanisms that are just beginning to be understood.

As frequently occurs in science, the insights on the mechanisms that explain a disease come from animal experiments. In this case, neonatology researchers from the Children's Hospital of Philadelphia showed that baby mice born to mothers that received antibiotics during pregnancy had a reduced immunological response. Similarly, a separate study showed that mice predisposed to diabetes and born to females that were given antibiotics had persistent alterations in their immune cells. These same mice developed diabetes a lot sooner than mice born to females that did not receive antibiotics. While a lot more research is still needed to fully understand all of this, it's becoming evident that complex interactions between microbes, the immune system, and other aspects of human metabolism, occurring as early as in utero (before birth), influence the risk of disease later in life.

Getting Smart About Antibiotics

In light of all these findings it is crucial to understand that using antibiotics should not be discouraged when they're really needed, but the overuse or abuse of antibiotics should be prevented. So, when are antibiotics necessary during pregnancy? The answer is simple: antibiotics should be taken for serious bacterial infections, and only bacterial infections. However, this can be hard to put into practice, especially during pregnancy, when doctors want to prevent any possible complications that may arise from an infection. Because of this, many health providers are too quick to prescribe antibiotics, as a safety precaution, to expectant mothers for ailments that don't require antibiotics, like the flu. The flu is a viral disease that causes symptoms that many people confuse for a bacterial respiratory infection. Its onset is very sudden and people feel awful for about a week, until they start getting better. It's not hard to imagine a pregnant woman showing up at a doctor's office almost begging to get a prescription that will make her feel a little bit better. However, antibiotics should not be used for the flu, regardless of how bad a patient feels.

There are exceptions to this, though; the flu can lead to secondary bacterial infections that do require antibiotic treatment. This usually manifests a little bit differently: you feel truly awful, and after a week or so, you start to get better, but then you start feeling worse, with coughing and chest congestion, which can lead to pneumonia. This is the classic example of a secondary bacterial infection following the flu, which should be treated with antibiotics.

However, the key concept here is to *prevent* infections from occurring in the first place if possible. As such, it is currently recommended that pregnant women get a flu shot. Fortunately we have an effective vaccine that is completely safe to use during pregnancy, which significantly decreases the chances of getting the flu and a secondary respiratory bacterial infection during flu season.

Despite the precautions you can take, infections do happen during pregnancy and antibiotics are prescribed. So what then? Based on the current research, it seems that the period at which antibiotics are taken is important, with microbial changes in the later stages of pregnancy being the most influential. If antibiotics must be used in the second and especially the third trimester, one should start or continue microbial supplementation with probiotics and a diet rich in fiber and vegetables. It's important to choose a probiotic that contains several species of *Lactobacillus* and *Bifidobacterium*, both known to be important early members of an infant's microbiota. As with any supplement or medication taken during pregnancy, we recommend discussing this with your health care provider.

Heading Off Group B Strep

During the births of her first two children, Neve had been given antibiotics, an increasingly common occurrence nowadays, with 1 in 3 women receiving antibiotics during labor. Neve knew how frequent antibiotic use is during delivery because she had tested positive for a type of bacteria known as Group B streptococcus, or GBS for her first two births. (Other very common circumstances that require antibiotics during labor are scheduled C-sections, which will be discussed extensively in chapter 4.) In many countries, all women between 35–37 weeks of gestation get tested for GBS. These bacteria commonly reside in 15–40 percent of all pregnant women, yet they rarely cause any symptoms. However, between 40–70 percent of GBS-positive women will pass it on to their babies during natural birth, and a small but very significant number of babies (1–2 percent) will develop a GBS infection (for further discussion of GBS infections, see chapter 4). Fortunately, if a pregnant woman who tests positive for GBS is treated with antibiotics during labor, the risk of her baby developing a GBS infection is reduced by 80 percent, making GBS prevention a pertinent use of antibiotics.

However, recent studies have shown that receiving antibiotics during labor alters the microbiota of the newborn, even if they are administered only an hour before birth. Reading about these studies made Neve, pregnant with her third child, feel uneasy. She knew that GBS could potentially be very serious and she understood the need for antibiotics during labor, but she wondered if anything could be done to *prevent* testing positive for GBS. Her second child has asthma and although it's impossible to know whether his exposure to antibiotics during birth is to blame, she's left wondering if it contributed. More importantly, Neve wanted to do whatever she could to decrease the risk of her new baby developing asthma, too. She hoped to help by testing negative for GBS, but how could she do something about that?

It turned out that she might actually have some say in the matter. GBS are bacteria that will expand in numbers only if they're given the chance. Normally other members of the microbiota keep them in check, usually our bacteria superstars, the *Lactobacilli* in the gut and the vagina. In fact, if you grow *Lactobacilli* and GBS together in the lab, the *Lactobacilli* make it very hard for GBS to multiply; they beat them easily. Furthermore, a small number of studies suggest that applying probiotics directly to the vagina increases *Lactobacilli* and decreases the number of GBS. This finding was shown in healthy nonpregnant women and remains to be supported in bigger studies, but given how safe it is to administer probiotics to pregnant women, Neve was open to trying this approach and her midwife supported this prophylactic treatment.

Neve ended up testing negative for GBS at her 36-week visit, and she is expecting to have an antibiotic-free birth very soon. However, it's important to mention that it remains to be proven in a randomized clinical trial that the prophylactic use of vaginal probiotics prevents or reduces the chance of a GBS-positive test during pregnancy. The use of vaginal probiotic suppositories, as with any treatments during pregnancy, should always be discussed with a health practitioner.

Can Bacteria Influence Us Before Birth?

So far we have discussed different ways to take care of the maternal microbiota during pregnancy in order to prepare the best kind of microbes that a mother can give to her baby at birth. This is when babies get soaked in microbes, during their trip down the vaginal canal. But very recent research shows that microbes may pay a visit to babies even before birth. For many years it has been widely accepted that humans are germ-free immediately before birth and that the presence of bacteria in utero is considered infectious and dangerous. Often this is true—bacteria growing in the placenta or the amniotic fluid can be a sign of infection and a cause of premature birth or even stillbirth. But what we're just now beginning to learn is that there may be very low numbers of bacteria that commonly reach the baby in the uterus without causing any harm. We still don't know how they get there and, more importantly, what they do, but in two separate studies bacteria were detected in the amniotic fluid and placentas of healthy babies. Although some scientists (including us) remain skeptical about these findings, the authors of these studies speculate that these bacteria are involved in immune stimulation of the fetus. Additional studies are needed before we can explain why this occurs, or if it even does.

Another more likely exposure to microbes before birth may occur in the form of bacterial metabolites, which are very small substances produced by the enormous amount of bacteria in our guts. Bacterial metabolites are known to travel in the bloodstream at all times, and are involved in biochemical reactions in just about every human organ, influencing many aspects of our metabolism. Thus, even if very few bacteria actually reach the fetus during pregnancy, the metabolites may reach the growing baby through the bloodstream and potentially affect fetal growth and development. Much-awaited studies are under way to explore the impact these microbes might have in human development before birth.

Dos and Don'ts

◆ **Do—** eat for your microbes, not just your cravings. Make vegetables, fruits, and fiber staples of your diet, along with the other food groups, and reduce sugary foods. A varied diet is a healthy diet for you, your baby, and your microbiota.

◆ **Do—** add daily probiotics, yogurt, or kefir (a fermented milk drink) to your diet. Increasing the growth of beneficial bacteria in your vagina will promote their passage to the newborn, where they carry out very important functions.

◆ **Do—** prevent infections if possible. Not only will you avoid feeling awful while pregnant, but it also reduces the chances of having to take antibiotics. Wash your hands often, avoid being in close contact with sick people, and follow the current recommendations of foods that pregnant women should avoid. If antibiotics are necessary, start or continue taking probiotics.

◆ **Don't—** sweat the small stuff, and do try to control stress as much as possible. Severe stress is associated with a number of disorders in children and also with alterations to the microbiome. If stress is becoming a big part of your life, reach out for help through your health practitioner. Even if your stress is moderate, incorporating exercise, yoga, or meditation into your routine can help keep the edge off.

◆ **Do—** consider vaginal probiotic suppositories in your third trimester in order to reduce the chances of testing positive for GBS. A negative GBS test will make an antibiotic-free birth more likely.

CARE FOR A SPOONFUL OF SOIL?

Perhaps the most bizarre of pregnancy cravings is the urge to eat dirt—a form of pica, a term used to describe an intense craving for nonfoods. Some suggest that dirt pica is the body's attempt to consume minerals and that it may be linked to iron deficiency, which occurs in many expectant women. Still, it is not known for certain what drives some mothers-to-be to eat dirt.

The rates of dirt pica vary depending on culture and socioeconomic status. In Kenya, it is so common that people see it as a sign of pregnancy, with 56 percent of pregnant women following this practice. Even in the US, 38 percent of low-income women from southern Mississippi claim to crave dirt or clay. Dirt pica is common enough that you can order dirt online to satisfy your craving! However, pregnant women are also more vulnerable to infectious diseases, and eating dirt may prove dangerous. Dirt is a known source of pathogens, toxins, and even lead, making it a bad option for those hard-to-curb cravings.

4: Birth: Welcome to the World of Microbes

The Best Laid Plans

At 3:50 a.m. a week before her due date, Elsa realized she was in labor. She was sleeping (sleeping should really have a different name in late pregnancy, as it is just not the same thing) when her water broke, alerting her and her startled husband that it was time. Soaking wet, they nervously laughed at the realization that they were going to meet their baby boy soon. They had a hospital delivery plan written down—labor in a bathtub, "laughing gas" for pain management, clear communication about interventions—and then, when the contractions became closer together, they would calmly put on comfortable clothes, gather their already-packed hospital bag (which included magazines, an iPad to serve as a music player and video camera, a massage device, and a heating pad), gather snacks and energy drinks, phone the grandparents, and drive to the hospital. The infant car seat had been installed in their car for about a month, and they had even practiced driving the route they were going to take. They already knew the best place to park in the hospital parking lot and the exact location of the maternity ward. Elsa and her husband had it all covered . . . or so they thought!

The first thing that kiboshed their perfect plan was having her water break before feeling contractions, also known as PROM (premature rupture of membranes). Elsa wanted to labor at home, but she knew that she had to go to the hospital right then. When the water breaks, the bag full of amniotic fluid, which keeps the baby protected, ruptures. It's not unusual for it to occur before labor, with 1 in 10 women experiencing that, but babies need to be monitored when this happens due to an increased risk in complications, such as an umbilical cord prolapse or an infection.

Within fifteen minutes they were out the door. They got dressed, grabbed the bag, forgot the snacks (oops), and decided to call their parents on the way to the hospital. It took Elsa another ten minutes to find a not-too-uncomfortable position to sit in the car, and just then, she started to feel her first real contraction. It was overwhelmingly strong. "If this is early labor," she thought, "I won't be able to deal with the pain." Elsa's husband, Paul, had previously volunteered to monitor her contractions. He had an app in his phone that would time contractions, and allow them to give each one an intensity score from one to five. As soon as Paul noticed the first contraction he reached for his phone and started to record its duration. Excited, he then asked Elsa: "How would you rate that contraction, babe?" With her gaze and voice lost, Elsa slowly opened her hand and showed him five fingers. "A five?" Paul said, "That can't be, we just got started!" And with the look that so many husbands have experienced during their wives' labor, Elsa just said, "Drive!"

By the time they reached the hospital, Elsa was already dilated five centimeters (halfway there) and in intense labor. "Forget the *&#^$ plan!!" she yelled. "I WANT AN EPIDURAL NOW!!" The nurse strapped a monitor to Elsa's belly to measure the baby's heart rate and Elsa's blood pressure. On the next contraction (they were coming three minutes apart now) the nurse noticed that the baby's heartbeat had dropped, not a lot, but enough to bring the obstetrician in to have a look. Then, just as the nurse was about to put an IV in Elsa's arm, the baby started squirming around, causing Elsa even more pain. Worse yet, the baby's heart rate dropped significantly. The obstetrician monitored the baby during the next sets of contractions and surmised that the baby must be pinching the umbilical cord. "We have to get him out *now*," the doctor said.

In what felt like hours but was only a few minutes, Elsa was rushed to the operating room and given spinal anesthesia for the C-section, after which they allowed Paul in the room. Elsa and Paul were both terrified.

However, very soon thereafter they heard the sweetest sound of their baby boy, Elijah, crying. A pediatrician and nurses quickly took Elijah to make sure he was all right (he was). After weighing and measuring him, they brought him to his parents, who were crying with relief, excitement, and love. "So much for the best laid plans," said Paul. Their cries turned into laughs as they realized that nothing had gone according to plan. It didn't matter . . . their baby was here and everyone was okay. Paul pulled out his phone, took the first picture of Elsa and Elijah, and sent it to the proud new grandparents, just over two hours after Elsa's water had broken, back in their bedroom.

Cesarean Epidemic

Although births come in different circumstances, durations, and outcomes, they have two things in common. First, just like with Elsa and Paul's experience, they seldom go as planned; births are unpredictable. Second, no one ever forgets when, how, and what it feels like to give birth. No other event in life compares in intensity and emotional impact. Biologically speaking, having a baby is the pinnacle of our existence, yet the human birth experience is very painful and often risky. In fact, compared to apes, human birth is longer and more perilous. Elsa's labor was unusually short at only two hours, but most first births average ten hours, and many are even longer. In addition, about 1 in 250 mothers carry a baby with a head too big to fit through the birth canal, requiring a cesarean section (C-section). One would think evolution would have favored easy deliveries, yet our bodies have not greatly improved on the process. Before the development of modern obstetrical medicine, there were about 70 deaths per 1,000 births. Those statistics have improved, but still, to this day, 500,000 women die annually worldwide from complications during childbirth. Why is human birth such hard and hazardous work?

Scientists believe that our births are more complicated because of the "human condition": we walk on two legs and have very big brains. Walking on two legs was truly advantageous to our human ancestors; they had their arms free to reach for fruit and other foods, they could carry items (babies included), they could hunt and craft tools, and they could look above the vegetation by standing upright. However, this advantage came with the anatomical price of narrower hips in order to achieve better balance and support the body's weight on two legs. Another aspect that makes humans unique is the large size of our brains. Thanks to our developed brains, humans can do math, build skyscrapers, and read books. Big brains (and, consequently, big heads) plus narrow hips? Any human can do this math: this causes the level five painful contractions Elsa was feeling and the medical need for C-sections.

C-sections are a medical miracle in terms of their ability to save the lives of so many mothers and babies. Try to imagine how much scarier Elsa's birth would have been had a C-section not been an option. Elijah's umbilical cord had twisted, preventing him from getting enough oxygen and blood flow. Elijah could have suffered a serious brain injury or even died from asphyxia if a trained doctor hadn't been able to pull him out surgically. A hundred years ago, dying during birth was a lot more common for both mothers and babies and modern C-sections played a pivotal role in changing this.

The history of when and where the first C-sections took place is a bit murky, but there are accounts of C-sections dating as far back as Ancient Greece. It is commonly believed that the name of this surgical procedure originates from the birth of the Roman emperor Julius Caesar. Regardless of whether this is true or not, Roman law decreed that all dying or dead birthing mothers had to be cut open in an attempt to save the child. Unfortunately, mothers rarely survived these early medical procedures and they were performed only as a last resort. Once anesthetic and antiseptic practices became the norm, C-sections became a much safer procedure and were used to save many lives. At the beginning of the twentieth century, for every 1,000 births, 9 women and 70 babies would die during childbirth, compared to 0.1 women and 7.2 babies today. That's more than a 90 percent reduction in mortality, a true triumph for modern medicine.

Still, for many decades C-sections were performed only when it was medically necessary: if the lives or health of the mother and/or the baby were at risk. However, towards the last quarter of the twentieth century, C-section rates skyrocketed. In 1970 the C-section rate was 5 percent in the US, rising to almost 25 percent by 1990 and to 33 percent in 2013. It has gone from a rate of 1 in 20 babies to 1 in 3 babies in the span of forty years. Canada's C-section rate is slightly lower at 27 percent, but it has still experienced a 45 percent increase since 1998.

Unlike the initial decrease in mother and infant mortality, the surge in C-section rates experienced in the past thirty-five years did not bring an improvement in mortality or morbidity (disease) rates. On the contrary, a C-section performed without a medical indication, also known as an elective C-section, is riskier than a vaginal birth. A C-section is a major surgical procedure that poses an increased risk of blood loss and infection for the mother. Also, any mother that has birthed via C-section can attest that healing takes much longer than a vaginal birth, not to mention the limited mobility of the new mother, who must let the incision to her abdomen heal; it's harder to hold the baby, to get up to change diapers (wait—maybe this is a plus), and sometimes even to breastfeed. Since 1985, the World Health Organization (WHO) has determined that the ideal rate for C-sections should be between 10–15 percent. Newer studies show that the number is likely closer to 10 percent. When C-sections rates approach 10 percent in a population, mortality surrounding birth decreases. But when the rates rise above 10 percent, mortality does not improve.

There are many explanations for this unnecessary but widespread increase in C-sections, and discussing them and their complexities are probably the subject for an entirely separate book. Suffice it to say, C-section rates are still increasing, and they are becoming epidemic and an emerging global health issue. Many experts disagree with this view and support the current rate of C-sections, because even if they are riskier than natural births, they are still very safe procedures. Modern obstetricians are extremely skilled in this surgery, and most complications that result from it, which are rare, can be treated with good outcomes in a hospital setting. There are maternal advantages associated with an elective C-section as well, including a reduction in urinary incontinence (loss of bladder control), avoidance of labor pain, reduction of fear and anxiety related to labor, and the overall convenience of planning the timing of birth. To some, the idea of a planned, painless birth is a dream come true.

On the baby front, C-section supporters claim that the health complications for babies born by elective C-section are rare and usually treatable. Babies born via C-section do look a bit different than babies born through the vaginal canal (their heads don't get squished), but after a few days they all look the same. However, while C-section advocates may be correct that severe birth complications, such as a stillbirth, are very rare in elective C-sections, we are now learning that there are significant health concerns associated with C-sections, including an increased risk of chronic disorders later in life, such as asthma, allergies, obesity, autism, IBD, and celiac disease. The elevated rates of these issues associated with C-sections hover around 20 percent for most of them. This is tremendously worrisome, considering that many countries have a C-section rate well above what the WHO recommends. Approximately 6.2 million unnecessary C-sections are performed around the world, with Brazil, China, the United States, Mexico, and Iran accounting for 75 percent of them. Brazil and China have an outright C-section epidemic; many hospitals in those countries deliver more than 85 percent of their babies surgically. The situation in Brazil has reached critical levels, as many women there have to give birth by C-section without the medical need for it, simply because of the shortage of hospital beds allotted for vaginal deliveries (see Brazilians Love C-sections, page 69).

The good news (kind of) is that it isn't the procedure itself that causes these disorders. Rather, it's something extremely important that *does not* occur during the few minutes it takes for a doctor to surgically remove a baby from the womb: the baby does not come in contact with his mother's microbe-rich vagina and feces.

A Dirty Birth Is a Good Birth

A baby's very first encounter with microbes most likely happens when his head comes out through his mother's vagina. As previously mentioned, the vagina contains an extremely high number of microbes, so the seconds (or minutes) it takes for a child to exit the birth canal are enough to impregnate a newborn's mouth, nose, eyes, and skin with many of them. It's also very common for women to defecate during birth, especially during the pushing stage. Babies usually exit the birth canal with their mouths facing their mom's anus, and it is now proposed that this position allows for additional exposure to maternal fecal microbes.

It makes total sense. The world is full of microbes, and all babies are going to get soaked with them immediately after birth, regardless of how they are born. Why not make sure that a baby gets coated in the microbes from which she will benefit most? Nature sees to it that the type of microbes first encountered by babies born vaginally are the ones that are going to aid in the digestion of milk, as well as contribute to the development of a baby's immature immune system, and even protect them against infections. Vaginal secretions are packed with *Lactobacillus*, whereas another milk-digesting bacteria known as *Bifidobacterium* come from feces. You've probably heard these two types of bacteria mentioned in yogurt advertisements. It's no coincidence that these bacteria are used in the dairy industry, as they're experts at digesting or fermenting milk and are also associated with health benefits. Unknowingly, every mother seeds her baby with a special custom package of microbes that will best suit her baby's needs. Babies instinctively seek their mother's breast shortly after birth, and breast milk is exactly what these microbes need to flourish in the baby's gut. This wonderful synchrony of biological events is a fine lesson in how nature works.

However, not every birth ensures the passage of beneficial microbes to newborns. As discussed in chapter 3, if the vaginal microbiota is unbalanced (low amounts of *Lactobacilli* in vaginal secretions), or if a woman has tested positive for Group B streptococcus (GBS), a baby will not get the same kind of microbial bath from her mom. Given how important it is to receive those beneficial microbes at birth, it's critical that women pay special attention to their vaginal microbiota in the weeks preceding birth. If there are any signs of a vaginal infection (itchiness, burning sensation during urination, or abnormal discharge), it's recommended that the mother consult a doctor and follow treatment with oral and vaginal probiotics as appropriate. In fact, given the proven safety of probiotics during pregnancy, all expectant mothers should consider including probiotics in their diet, especially in the weeks preceding birth (see additional recommendations in chapter 3).

If one could view birth through a microscope, a C-section is drastically different than a vaginal delivery: their microbiota is remarkably dissimilar. Studies comparing the gut microbiota of newborns in the days and weeks following birth consistently show that babies born by C-section have lower numbers of *Lactobacillus* and *Bifidobacterium*, as well as divergences in several other bacteria. These babies are colonized by microbes often found on skin, soil, and other external surfaces, instead of vaginal and fecal microbes. Even more worrisome, some of these differences persist and can still be detected when children are seven years old, according to a 2014 Dutch study.

To better understand how different a C-section is in the context of microbes, lets trace a baby's possible route of microbial exposure following a C-section. The brand-new bundle of joy goes from the doctor's sterile gloved hands to a table or a scale where he's touched with medical utensils and cloths. He may also brush someone's lab coat or hand in the process. If all is well, minutes later the baby is brought to his parents, and they can finally touch and kiss him, providing skin and mouth contact. Very often the baby is not allowed to breastfeed until his mother has started to recover from the anesthesia, which takes hours in most cases (although a few hospitals are now allowing this right after delivery). During this period, the baby will likely be wiped clean, warmly bundled in a clean hospital blanket, and placed in a cot, heated by a lamp, where he is offered warm (sterile) formula. During all this, the baby is exposed to the air, which has many microbes, but they are very different from mom's microbes, the ones humans are adapted to get exposed to at birth. It can take up to two hours before the baby is returned to his mother, when he can finally try breastfeeding for the first time.

Seeding Hope for the Future

Clearly, a baby born via C-section surely misses out on something crucial: that first splash of mom's microbes. But rather than judging mothers who have decided to give birth this way, whether by choice or due to medical necessity, we need to look at what can be done to make C-sections a more microbiota-friendly choice.

How can one restore a baby's microbiota following a C-section?

If you think about it, the way vaginally born babies are exposed to microbes is very simple: they come in contact with vaginal secretions. Why not inoculate a baby born by C-section with mom's vaginal secretions shortly after birth? Such procedures, called "seeding," are currently being used and tested in several hospitals around the world, and have been gaining an increasing amount of attention.

Veronica, a thirty-three-year-old mom from Edmonton, Canada, had to schedule a C-section some weeks prior to her due date because her baby was in breech position. However, she was aware of the importance of imparting her microbiota to her baby during vaginal birth and decided to talk to her midwife about this. Her midwife came up with a plan. She inserted a piece of sterile gauze into Veronica's vagina while she was waiting to be taken to the operating room. Minutes before her C-section, her midwife removed the gauze and placed it in a sterile glass container. Right after their baby girl was born, Veronica's husband took the gauze with gloved hands and swabbed it inside the baby's mouth and on her skin. Veronica also swabbed her own nipples, with the hope that the infant would take in even more vaginal microbes while breastfeeding.

As far-fetched as this method may sound, Veronica is part of a growing trend of moms and health practitioners who are trying it. Not only does it make scientific sense, but there's also scientific evidence backing up its effectiveness. Dr. Maria Dominguez-Bello, a scientist at NYU and a leading expert in the field of microbiota studies, has focused her attention on the development of early microbiota. She recently conducted a study involving eighteen births, in which babies born by C-section were "seeded" with mom's vaginal secretions and placed on mom's chest. Her team found that this process resulted in the microbiota of "seeded" babies becoming much more similar to that of a baby born vaginally. "While not equivalent to a baby born vaginally, there is some important restoration happening," she says. It's still unknown whether this simple procedure will reduce a baby's risk of suffering a chronic illness later in life. Her research group will follow up with these children in the years to come. Additionally, her group is working on conducting a much larger study that can provide sufficient evidence in terms of the safety of this practice. In the meantime, there's a compelling argument that women planning to have a C-section should discuss this option with their doctor or midwife.

Antibiotics During Birth

Antibiotics are routinely administered in conjunction with a C-section, given intravenously as a precaution against infection. As one can imagine, with the surge in C-sections, there has been a similar increase in the use of antibiotics during birth. In this instance, the antibiotics are truly necessary, as 10–15 percent of women that undergo C-sections will develop an infection. But it's up for debate whether the antibiotics have to be administered before surgery, or if it can wait until after the baby has been delivered. If given before the C-section, the baby will likely be exposed to the antibiotics, further compromising her microbiota at birth. If given after, the mother will still get the treatment she needs to prevent an infection and the baby will not be directly exposed to the antibiotic.

This was the case for Carley, now the mom of a healthy three-month-old daughter. During a doctor visit early in her third trimester, Carley learned she would have to deliver her baby via C-section (an umbilical cord abnormality made a vaginal birth too risky). As a naturopathic doctor herself, Carley had hoped for a vaginal birth, but she was aware of the need for a C-section for the safety of both her and her baby in this case. At the same time, Carley was aware that C-section babies have an increased risk of developing allergies, asthma, and obesity, with current research showing that a difference in microbial exposure influenced this risk. She had been taking daily probiotics throughout her pregnancy, but knowing that she would receive antibiotics before her birth, she was concerned that her baby would not received the optimal amount and type of microbes during birth. Carley explained her concerns to her obstetrician, who agreed to administer the antibiotics after her baby was born. They also agreed to "seed" her baby with her vaginal secretions after birth. Carley's C-section went smoothly and she recovered very well from it. She continued to take probiotics and to eat a healthy and varied diet to help restore her microbiota afterwards.

As in Carley's case, doctors are getting an increasing number of requests to administer antibiotics to the mother only after the baby is delivered, and even to forego antibiotic treatment altogether. While delaying the administration of antibiotics is a reasonable proposition, eliminating antibiotics during a major surgical procedure puts the mother at a very significant risk of infection. Like all medical decisions, the risks must not outweigh a patient's benefits. In this case, the desire to protect the mother's microbiota is outweighed by the increased risk of a severe infection acquired during surgery.

Another common use of antibiotics at birth is the application of antibiotic ointment (erythromycin) in the eyes of newborns. This is routine in the US and Canada, aimed at preventing the development of eye infections from the bacteria that cause gonorrhea and blindness caused by chlamydia. Because the possible outcome of these infections in a newborn is so severe, it is a medical indication in all births, although countries such as Australia, the UK, Norway, and Sweden forego the practice. In the US, thirty-two states are required by law to administer this treatment, regardless of whether the mother has chlamydia or gonorrhea, or whether the baby was born vaginally or via C-section (the infection can occur only during a vaginal birth). Recently, the Canadian Paediatric Society stopped recommending routine eye prophylaxis; however, this has not yet filtered down to common practice and many children still receive this treatment.

All pregnant women should be tested for sexually transmitted infections (STIs), including chlamydia and gonorrhea, and in fact most of them already are. But considering that the majority of women test negative for these diseases, and that many of them are part of a monogamous relationship, it seems reasonable to recommend eliminating the use of topical erythromycin after birth, at least in places where the law allows for a parent's right for an informed refusal of treatment. Although a small amount of antibiotic in the eyes will not have the same effect as an antibiotic administered intravenously, it could certainly affect the microbiota on the skin. Additionally, indiscriminate use of antibiotics aids in the development of antibiotic resistance, a larger public health issue.

Several other circumstances require the use of antibiotics during birth, including premature water breaking, labor lasting for more than twenty-four hours, signs of infection (e.g., fever) in the mother or the newborn, or if the mother has a known infection (such as a urinary tract infection). After birth, if the baby shows any symptoms that could indicate infection, the baby is tested. This usually takes 24–48 hours, during which it is assumed that an infection is taking place and the baby is administered antibiotics while awaiting laboratory results. In the vast majority of cases, these tests turn out to be negative, meaning many babies are given antibiotics unnecessarily. However, the consequences of undertreating an infant that is indeed suffering from a life-threatening infection can be disastrous. Clearly, there's a real need for better and faster methods to diagnose newborn infections, but until then antibiotic use in these circumstances is medically necessary. Newborns are especially susceptible to diseases, given how immature their immune systems are at birth, and the outcome of an infection can be very severe. Antibiotics certainly have a place during and after birth, and they have saved many lives, but considering how strongly they affect a baby's microbiota, their use should be limited to medical necessities.

Premature Babies

Some pregnancy complications can lead to the delivery of a baby well before she is ready to survive outside the womb. Medical treatments have advanced enormously in this field and premature babies can sometimes survive when born as early as twenty-three weeks (barely five months of pregnancy!). These babies often face major difficulties, like the inability to breathe or eat on their own, and much of their development has to occur in a hospital incubator, under the vigilant care of doctors and nurses in neonatal intensive care units (NICUs). Despite best efforts, this just isn't the same as a dark, warm, wet womb. Premature babies are born with immature intestines, making them vulnerable to the development of an extremely serious intestinal disease known as necrotizing enterocolitis, or NEC. Around 7 percent of low birth weight infants suffer from NEC in the US and Canada, and up to 30 percent of these infants die as a result. Naturally, there has been a big push to determine what causes NEC and how to prevent it.

NEC usually occurs a week or more after birth, and pathogenic bacteria (bacteria that cause disease) are often found in premature babies that suffer from NEC. This has led to the suspicion that NEC is caused by an infectious agent. Recent microbiota surveys from these babies have shown that NEC is very likely caused by an imbalance of the immature microbiota in their young bodies. Researchers found that in healthy premature babies that do not develop NEC, their microbiota becomes similar to a full-term baby at around six weeks of age. In contrast, premature babies that develop NEC have an abnormal growth of harmful bacteria days or even weeks before the symptoms of NEC appear.

The microbiota of premature babies have a lot going against them. Since these children are frequently delivered by C-section and given antibiotics after birth even when they're not showing any signs of infections, it's no surprise that a balanced bacterial intestinal community cannot develop properly in their bodies. However, recent studies have shown that the incidence of NEC can be significantly reduced if these babies start probiotic therapy shortly after birth. In a large study of three hundred premature babies in the NICU of a large hospital in Montreal, the administration of a mixture of *Bifidobacterium* and *Lactobacillus* led to a 50 percent reduction in NEC. That's a pretty astonishing figure! The authors of the study estimated that about twenty-five hundred cases of NEC in North America could be prevented each year by this very simple treatment.

Dos and Don'ts

♦ **Do**— your homework regarding your chosen mode of delivery. There are a number of studies that can inform a pregnant woman about the risks and benefits of vaginal birth and C-sections. Include what is known about the microbiota during birth as part of your decision. Also remember that births usually don't go as planned, so make yourself knowledgeable in case you need to make a quick but necessary change in plans.

◆ **Do—** look after your vaginal and gut microbiota during the weeks preceding birth (see chapter 3). A high number of *Lactobacillus* and *Bifidobacterium* will help establish these microbes in your baby, and help him adjust to life outside the womb. Take probiotics containing combinations of several of these microbes to boost their levels in your body.

◆ **Do—** discuss with your midwife or physician the procedure known as "seeding," to help restore your baby's microbiota if she is delivered by C-section.

◆ **Do—** consider sharing your concerns regarding the use of antibiotics during birth with your midwife or physician. If medically unnecessary, antibiotics should not be given as a routine measure to either the mother or the baby.

◆ **Don't—** question the need for antibiotics if medically necessary. If mother and/or baby are deemed at risk of infection during or after birth, antibiotics are the only effective treatment to prevent a severe outcome. Trust your health care practitioner. If a baby has been given antibiotics shortly after birth, ask your doctor if he can be given infant probiotics soon after the treatment stops.

◆ **Do—** talk to the pediatrician about the use of probiotic treatment, if you have a premature baby who has to spend time in a NICU. Probiotic use significantly reduces the risk of developing a very severe disease known as necrotizing enterocolitis (NEC).

BRAZILIANS LOVE C-SECTIONS

No other country in the world beats Brazil for number of C-sections. A large proportion of Brazil's growing middle class use private medical insurance, which means that certain private hospitals look after these patients exclusively. In many of these hospitals, obstetricians rarely attempt natural births, and the rates of C-sections reach 90 percent. Because of this, a C-section birth is now the norm for millions of middle class and affluent Brazilians, and has become a type of status symbol.

Even public hospitals have succumbed to this trend, with almost 50 percent of all births in Brazil delivered by C-section. With so many women booked for C-sections in advance, it has become difficult to find available hospital beds for vaginal births, which take much longer than scheduled C-sections. Some physicians are known to ask for extra payment if a patient wants to deliver naturally, as it involves many more hours of medical care (and often occurs in the middle of the night).

This issue has become so critical that the Ministry of Health recently approved new rules aimed at reducing Brazil's alarming C-section rate. Doctors are now mandated to inform women about the risks of C-sections and ask them to give their signed consent before the procedure. Additionally, doctors have to medically justify the need for a C-section based on a complete record of labor and birth. It is not yet known whether these rules will improve a Brazilian woman's chance to give birth vaginally if she wants to.

5: Breast Milk: Liquid Gold

Born Too Young

After minutes (or hours!) of intense pushing or skillful surgical maneuvers, a brand-new life meets the world. What a beautiful moment! Finally, after months of pregnancy and the intensity of the actual birth, all the hard work is over and parents can enjoy touching and smelling their new child. Ha! As if! For most parents, it becomes evident very quickly that the hard work is just beginning: sleepless nights coupled with the near-constant needs of a newborn.

Babies demand a lot of care during those initial few months. Compared to most other mammals, humans are born in a very immature and fragile state. While in utero, all babies grow to a point in their development in which they can withstand being outside of the womb. For example, a calf or a foal will shakily stand up and take its first steps mere hours after birth. In contrast, a human mother needs to deliver her baby before his head becomes too big to fit through the birth canal, which necessitates babies being born somewhat immaturely. For humans, it takes an additional 5–12 months for a baby to move on its own and eat anything other than milk.

During this somewhat fragile early stage of life, a baby's developing intestines and other organs demand constant nutrition and a lot of rest. This translates into feeding every 2–4 hours and sleeping around the clock, hence the unmistakable zombie-like look most parents have during the first months after birth.

A baby's intestine in particular goes through a lot after birth. Not only does the newborn gut have to mature and develop all the mechanisms necessary to digest and absorb food, it also has to withstand the onslaught of trillions of foreign microbial cells rapidly colonizing its entire surface. On one hand, a baby's gut has to allow the entry of the much-needed nutrients from milk, but it also has to prevent microbes from entering the rest of the body. This is an important concept to grasp: while trillions of microbes inhabit our intestines, they live in the inner intestinal space, known as the intestinal lumen. Think of the gut as a tube (which it is; see chapter 12), with the mouth and the anus at either end of that tube, and the rest of our body surrounding it. The intestinal lumen is the inner part of the tube, and although trillions of bugs live and perform all sorts of good tasks for us in the lumen, they aren't supposed to breach the borders of that tube. Our internal organs, such as the heart or the kidneys, do not benefit from having microbes floating around them. Quite the contrary—the entry of microbes into our body is a signal for our immune system to react, attack, and clear the intruders, and a strong immune response at this stage of life carries too much risk in such a wee little body. Thus, keeping microbes at bay within the intestinal lumen is a good strategy to prevent infection and a strong immune reaction. As mentioned in chapter 2, an overly active immune system can create havoc, so a calm and tolerant immune state is favored in the presence of so many microbes.

With all of this going on, the developing gut of a newborn is a somewhat chaotic place for a few months. It's no surprise that babies seem uncomfortable right after they eat, needing to be burped and sometimes shocking parents with the loud and messy sounds that come out of both ends. This is one of the main reasons why we can't go ahead and feed pasta with meatballs to a baby soon after he's born. It takes about 4–6 months for his intestines and immune system to be ready for all the nutritious foods you want to give him. Yet as usually happens in nature, millions of years of evolution have designed a perfect food for this challenging developmental process: human milk.

Feeding Trillions

Health organizations around the world agree that human breast milk is an amazing liquid and that it's the healthiest food for babies. Its benefits are most noticeable when babies drink it exclusively during the first 4–6 months and combined with solid food until age two. Scientists have studied breast milk for a long time, yet we're still discovering fascinating facts hidden in its biochemical composition, which is extremely complex and actually changes depending on many factors, such as how far along in the pregnancy birth occurs, the baby's age, and the woman's diet.

The nutritional components of breast milk include proteins (~10 percent), fat (~30 percent), and carbohydrates (~60 percent), plus many vitamins and other small molecules, encompassing all the necessary nutrients a baby needs in order to grow. As anyone who has raised a baby knows, they grow extremely quickly. It's not your imagination when you notice that your child grew overnight; babies grow every single day, hence the constant need for food and a mother's constant breast engorgement and accompanying exhaustion.

In addition to the nutritional components in breast milk, there is an ever-expanding list of ingredients found in it that are extremely beneficial for a baby's development. For example, maternal antibodies help fight potential pathogens (disease-causing microbes) that may enter a baby's immature gut; *lactoferrin*, a protein that binds iron, steals it from iron-loving bacteria (often pathogens) and prevents them from thriving; *lysozyme*, a very potent enzyme, aids in food digestion; and *growth factors*, which are potent immunity-enhancing substances, promote intestinal development and at the same time keep the immune system tolerant. All of these are present in breast milk (although not in formula), and they help the intestine mature while protecting it from the overwhelming load of microbes rapidly setting up house.

Interestingly, not all nutrients in breast milk can be digested by the newborn's intestinal system. A significant amount of the breast milk carbohydrates—known as *oligosaccharides*—go right through a baby's stomach and small intestine without being digested by a baby's digestive enzymes. Ready for the fun and fascinating part? It turns out that oligosaccharides, which comprise about 10 percent of human milk content, are *only* digested by bacteria present in the baby's large intestine. Hence, a nursing mom is not just feeding her baby's cells, she is devoting about 10 percent of her breast milk to feed the trillions of bacteria that have colonized her baby's gut.

It takes a lot of calories to make breast milk. Why would a mother's body invest so much time and energy to produce and pack into breast milk a certain nutrient exclusively to feed her baby's bacteria?

The only explanation is that the baby benefits immensely from hosting these bacteria and mom wants to make sure they're well watered and fed. These bacteria do a lot of things in those early weeks and months that affect the infant. It's been shown that the metabolic activity of a baby's microbiota (that is, breaking down stuff and producing energy) is even higher than that of the liver.

Recent research also shows that many of the benefits historically attributed to breast milk are actually mediated by the baby's microbiota. For example, the presence of a specific *Bifidobacteria* species enhances the effect of a key growth factor present in breast milk, known as the transforming growth factor, which keeps the immune system tolerant, thereby preventing an excessive inflammatory response in the gut. Administering *Bifidobacterium breve* to preterm infants made them more responsive to the effects of this growth factor, thus showing that the presence of this bacteria is an important component in early development.

Yet there's more: Feeding a baby's microbiota is not the only way that breast milk promotes the growth of beneficial microbes. Cutting-edge research has recently shown that breast milk itself comes with its own microbiota. These bacteria, as we've discovered, come not only from the mother's skin (when a child breastfeeds, a lot of skin bacteria make it into his tummy), but from the milk itself. At first these findings puzzled some scientists, who doubted the results and assumed that the breast milk samples were contaminated during the sampling process. It took many experiments to show that breast milk indeed has its own bacterial residents. Before these findings came to light, breast milk was thought to be a sterile fluid, and the presence of bacteria was a sign of a painful breast infection known as mastitis.

In one of the experiments, pregnant mice were orally given specific

Lactobacillus bacteria, which were labeled in a way that would distinguish them from other bacteria and that could also be detected later. Lo and behold, the scientists found that the exact bacteria they had fed to the pregnant mice turned up both in the mice's breast milk and in the baby mice's tummies. The research group, from Complutense University of Madrid, later went on to show that the same was true in humans. Somehow, a mother's body manages to add her own bacteria to her breast milk in order to further promote the growth of beneficial bacteria in her baby, just like packing a special treat in a lunch box.

How does this happen? Where are these bacteria coming from? The truth is, we're still not sure how this happens, but recent experiments have shown that these bacteria may come from the mother's gut. Specialized immune cells that live in the intestines protrude out of the intestinal wall, reaching into the intestinal lumen to "swallow" bacteria. Some scientists hypothesize that these cells then take a long ride from the intestines to the mammary glands, where they're passed to the nursing baby, along with their bacterial hostages. Others propose that the bacteria themselves sneak out of the intestinal lumen and make their own way to the breast, without the need of an immune cell as an escort. This process marks yet another way breast milk shapes the bacterial ecosystem of a baby's gut.

Regardless of how they get there, bacteria inhabit breast milk—and there are a lot of them! A detailed analysis of the type of bacteria that live in breast milk found that they're present in surprisingly large amounts. A single feeding can provide a baby with up to 100,000 bacteria. This analysis, performed in Dr. Mark McGuire's lab at the University of Idaho, also showed that the breast milk microbiota doesn't contain just a few species of bacteria, but is quite a diverse community. Even colostrum (the first watery fluid produced by the breast right after delivery) has hundreds of types of bacteria. What's even more interesting is that the type of bacteria changes depending on several factors, such as the age of the infant and even the mode of delivery. For example, the type of microbiota in colostrum is very different from the microbiota in breast milk at one or six months after birth. Also, the breast milk microbiota in mothers that gave birth vaginally is different than the breast milk microbiota in women following a C-section. It's not known why this is, but it's speculated that the physiological stress and hormonal changes that happen during vaginal birth influence the transmission of microbes into the breast. Clearly, breastfeeding is not the sterile practice many people might think it is. Every single time a baby reaches for a breast to feed, he's not only getting calories, he's also acquiring mouthfuls of beneficial bacteria and the right food to keep these bacteria fed.

Breastfeeding: Not as Easy as It Sounds

Pregnant women and new mothers around the world hear time and again that breastfeeding is the best food for the baby, that it promotes emotional attachment between mother and child, that it provides all the nutrition a baby needs, not to mention that it also helps burn all those extra pounds gained during pregnancy. It seems that every time a mother-to-be sits down in a doctor's office there's a brochure featuring a picture of a blissful-looking mother nurturing her baby from her breast. *Breast is best* is the word on the street, and it truly is, but it's rare to hear the truth about how hard and exhausting breastfeeding actually is, especially in the beginning.

Jacky, the mom of six-year-old Steph, vividly remembers her agonizing first few weeks of motherhood. It didn't matter how hard she tried, Steph wouldn't latch properly. Steph would also only remain calm or fall asleep when she was on her mom's breast, always waking up the second Jacky tried to pull her off. By the time Steph was a week old, Jacky couldn't handle it anymore. Every time Steph would latch, it was unbearably painful; the delicate skin covering her nipples was split open in many places, and she would bleed during every feeding, a truly torturous experience Jacky will never forget. The advice she was getting from her mother and aunt was not what she wanted to hear: "Give her formula!" they kept saying. "You were fed formula when you were a baby and you were a healthy kid." They were right, she was a superhealthy kid, but she really struggled with the idea of not providing breast milk to her baby given everything she knew about its benefits.

Jacky was very close to giving up on breastfeeding when even the weight of her clothes on her breasts was so painful that she would break into tears. Instead, she decided to listen to a friend, who recommended a lactation consultant, and the following day, Jacky, her husband, and Steph found themselves in the overcrowded waiting room of this particular professional. The room was filled with overwhelmed and exhausted moms, who, just like her, were desperate to figure out how to make breastfeeding work. Yet there they were again—the stupid brochures with a picture of a beautiful and surprisingly rested mom blissfully breastfeeding her child. Jacky wanted to burn every single one of them. Instead she waited patiently, hoping that help was on its way. It took quite a bit of work and several visits to the consultant, but eventually Steph learned to latch correctly and breastfeeding became a lot easier and painless.

Still, Jacky couldn't help but wonder why no one had mentioned how hard breastfeeding was going to be. Can the proven benefits of breastfeeding really outweigh the risks of parenting while emotionally spent? There's no doubt that breastfeeding is worth the effort a new mother puts into it, but the process is not always as intuitive or easy as it may look, and a very large number of women don't receive proper advice on how to do it, or are unable to breast-feed for a variety of reasons.

The statistics don't lie: Six out of ten moms experience difficulty breastfeeding during the first six months, and three out of ten suffer mastitis, a painful and debilitating infection caused by pathogenic bacteria that make their way into the breast. These bacteria infect one or more mammary glands, causing redness, swelling, fever, and pain—quite a bit of it. Women are instructed to continue breast-feeding or to express their milk to help wash out the bacteria, and they are often also prescribed antibiotics. Although a necessary mea-sure, breastfeeding while taking antibiotics to treat a breast infec-tion will undoubtedly result in the baby drinking antibiotic-laden milk—not at all the cocktail the baby's microbiota wants to drink. Antibiotics will also likely change the microbiota of the breast milk, further altering the type of bacteria that a mom is passing on to her baby. In this situation, it's very advisable to administer a high dose of probiotics to the mom, and also to give pediatric probiotics to the baby to replenish at least some of the key beneficial species of bacte-ria (these probiotics should contain a mixture of different *Lactobacilli* and *Bifidobacteria* species).

A serendipitous discovery made by a group of scientists research-ing the microbiota in breast milk suggests that probiotics can be used as a treatment for mastitis as well. They found that when they gave a mixture of *Lactobacilli* species to nursing moms, those who were suffering from mastitis experienced a noticeable improvement. In a different study, 352 moms with mastitis were treated orally either with antibiotics or with two types of *Lactobacilli* probiotics. Within three weeks of receiving treatment, the women given probiotics experienced much lower pain scores, and 88 percent of them fully recovered, compared to 29 percent of the women given antibiotics. Recurrence of mastitis was also much lower in the probiotic group, with only 10 percent of those women developing mastitis again, compared to 30 percent of the antibiotics group. This was a pretty surprising finding, until one realizes that certain *Lactobacilli* species are quite adept at fighting off disease-causing bacteria. Based on these studies, women experiencing symptoms of mastitis might consider using probiotics as a treatment. However, it's critically important to recognize that mastitis can be a very dangerous infection and that medical advice should be sought out and followed as well.

When Breast Milk Is Not an Option

Millions of babies around the world drink formula instead of breast milk for much of their infant and toddler lives. Baby formula is a societal need, because the realities surrounding the decision or the length of time a woman breastfeeds is contingent on many factors, including cultural and religious practices, as well as socioeconomic reasons.

Formula is a helpful and often necessary solution for so many women, and it does contain the nutrients that babies need to grow and thrive. A significant number of women in the world don't produce enough milk (if not expressed consistently, breast milk production can dwindle, even more so during the early postpartum period). Other women may pass on an infection (e.g., HIV) or a drug to their babies, making formula a safer option. Some women live and work in countries where there are limited or even nonexistent maternity work benefits and they simply cannot afford the time to breastfeed. Others opt for formula because of the demands of their careers. Regardless of the reason, the decision to breastfeed a child is entirely personal, and ideally, an informed one. Not breastfeeding a child does not equal inferior parenting; it's simply a different approach to nourishing a child.

Certainly, breastfeeding should be promoted and societal changes are needed to make breast milk available to more children, but until that happens, science should also look at improving baby formulas to mimic the incredibly complex composition of breast milk. Baby formulas have seen significant improvement over the years, but we still lack the scientific and technical knowledge to make formula exactly the same as breast milk—that will take much more time, if it ever happens. Until then, one way formula can come closer to breast milk is by adding microbes to it in the form of probiotics.

Only very recently has the scientific community realized the role of the microbiota within breast milk. One constant in studies on the topic is that babies fed with formula have a very different microbiota than babies who are breastfed. How different? It's been reported that breastfed infants tend to have a more uniform population in the gut. There are also differences in the type of *Bifidobacteria* that live in the gut, as well as in many of the other less abundant species. There are also distinctions in the type of bacterial metabolites present in breast milk and formula, suggesting that the various bacteria in breast milk do very different things. Also, formulas contain cow-derived oligosaccharides, not human oligosaccharides, further altering the type of food available for the microbiota of breastfed children.

This area of research is incredibly new, and very little is known about the changes that these differences create at the molecular level. However, the epidemiological data that link formula feeding with a variety of diseases is, in some cases, very strong. Formula feeding is associated with a higher risk of developing infectious diseases, obesity, and asthma, and the evident changes in the microbiota hint at the involvement of early-life microbes in these diseases.

What's more important is that a number of studies have shown that adding probiotics to formula makes a formula-fed infant's microbiota more similar to that of breastfed babies. Other ways that probiotic administration brings formula closer to breast milk is by improving stool consistency and frequency. Because most studies suggest beneficial clinical effects, and because we know that probiotics and prebiotics (foods that promote the growth of probiotic species; see chapter 16) are very safe, it makes sense that formulas should include them, and many formula brands already do. In addition to choosing a formula with probiotics, parents may further supplement with probiotic drops.

Breastfeeding After a C-section

Whether scheduled or not, a C-section makes breastfeeding an even bigger challenge. Melanie, mom to three babies born this way, remembers well how difficult it was for her to breastfeed after her first C-section, despite her best intentions to do so. She opted for a C-section upon advice from her doctor, since her baby remained in breech position towards the end of her pregnancy. Disappointed at first, Melanie became comfortable with the idea of a surgical birth and was even relieved knowing that she would avoid labor and pain (who wouldn't?).

A few hours after she was admitted for her C-section, Melanie began to realize that foregoing labor did not mean her birth was going to be easy. In fact, she was incredibly anxious during the procedure, as well as uneasy from feeling the forceful (though mostly painless) pushing and pulling movements inside her abdomen. Things definitely got trickier after birth. Recovering from the effects of the epidural anesthesia proved to be really hard for her. She was dizzy and nauseous and her legs and arms were extremely numb; even three hours after delivery she wasn't comfortable holding her baby out of fear that she would drop her. For days afterwards Melanie felt like she had been run over by a train, and she struggled to cope with the early days of parenting, and with breastfeeding. Melanie's abdomen remained sore for weeks, and she still couldn't move well or hold her baby comfortably the first few days after birth.

Breastfeeding was difficult from the get-go for Melanie. She had a brief skin-to-skin moment with her child while she was being sutured, but this lasted only a couple of minutes and it was really more of a face-to-face, as she was unable to actually hold her baby. For the next few hours, her infant remained in the nursery, where she was given warm formula to keep her nourished and comfortable. By the time Melanie was reunited with her little girl, she was peacefully sleeping in her cot and continued to sleep for a few more hours. The first time Melanie was able to offer her breast was almost six hours after birth, and it was with the help of her husband, as she didn't feel safe holding her baby yet. Throughout her stay at the hospital, Melanie continued to breastfeed through the nausea, dizziness, and pain. She needed assistance, as she couldn't stand up for the first twenty-four hours, so it was her husband or a nurse who brought her baby to her and held her while she nursed.

Once they got home it became even harder. Although Melanie was taking pain medication, her incision caused her a lot of discomfort, and getting up to pick up and feed her baby was excruciating.

This was taking a toll on her husband as well, who hadn't had much sleep for days and was very frustrated by not being able to ease his wife's pain or his baby's constant hunger. They gave their infant a few more formula feeds at night to allow Melanie to rest and recover. By the fourth day, Melanie's milk came and with it a new challenge: more pain and constant engorgement. Her little girl wanted to breastfed every two hours, as most newborns do, but it was incredibly difficult for Melanie to find a comfortable position; it just hurt too much. By the end of the first week, Melanie was completely exhausted, so they decided to switch to formula at nights to give Melanie much needed rest. It worked; her baby started sleeping for 3–4 hours at night and that felt like heaven for Melanie. However, her baby began rejecting her breast and preferring her bottle, a common occurrence known as "nipple confusion." It's easier for a baby to drink from a bottle than from a breast, and many of them choose the bottle when given the option. By the time Melanie's baby was a month old, she was feeding only once or twice from the breast and taking the rest of her nutrition from formula. At three months old, she was taking only formula; Melanie's milk had dried out and her baby rarely accepted it anyway.

Melanie's first baby had very few issues with formula. She grew beautifully, gaining weight as she was supposed to, and overall was a happy and healthy child. Still, when Melanie got pregnant again, she decided to give breastfeeding another chance based on all the things she read about breast milk being the best nourishment for babies. Aware that she would have another C-section (following a C-section with a vaginal birth is not impossible, but most health care practitioners advise against it due to an increased chance of uterine rupture during labor), she knew that breastfeeding might be challenging again. She talked to her doctor and explained that she wanted to breastfeed her baby right after birth. So instead of offering formula, she enrolled the help of a nurse, who brought her baby to her to breastfeed while she was recovering from the anesthesia. Then her mother-in-law stayed with her during the night to help handle the baby during feedings. Once they got home, friends provided support—looking after their older child (now two years old) and doing some household chores. Melanie knew that if she wanted to succeed, she needed the time to spend with her baby, as well as the time to recover from a major surgery. It wasn't easy, but Melanie managed to exclusively breastfeed her two younger children. Her advice for mothers who want to breastfeed after a C-section? *"Be determined and get help!"* (See further recommendations in Best Bets for Breastfeeding, page 86).

Dos and Don'ts

- ◆ **Do**— get informed about the benefits of breastfeeding. Breast milk not only provides the best type of nutrition to babies, it also contains the right type of nutrients for a baby's microbiota. Breastfeeding has been repeatedly associated with better health outcomes, including protection against infections, asthma, and obesity.

- ◆ **Don't**— expect breastfeeding to be easy, because often it is not, at least during the first few weeks. Prepare for it as you prepare for birth by attending a prenatal clinic or getting advice from a nurse, doula, or lactation consultant. Be patient, as breastfeeding correctly is not always as intuitive as it looks and it takes time to do it correctly. If you experience difficulties breastfeeding, look for help as soon as possible.

The first few weeks after birth are both overwhelming and very important to establish successful breastfeeding, so this is the right time to ask for professional help.

◆ **Do—** consider taking probiotics if mastitis or other infections develop, especially if they require antibiotics. Probiotics will help replenish the microbes that inhabit breast milk, which may have been lost due to antibiotic treatment. Probiotics have also been shown to be effective in treating mastitis, so consider using them as an alternative form of treatment for this condition—however, always consult with your health care practitioner first.

◆ **Don't—** think that providing formula to your baby is an inferior form of parenting. Formula is a societal need and the decision to breastfeed or not ultimately depends on the mother. If you decide to use formula, complement it with pediatric probiotics. One of the biggest differences between breast milk and formula is the type of microbiota that result from them, so probiotics will help make formula more similar to breast milk. Consult your doctor or nutritionist for advice on the best options in the market.

◆ **Do—** look for extra help after a C-section. Breastfeeding is hard as it is, without the added physical and mental recovery from major abdominal surgery. Establishing successful breastfeeding will depend on preparing for it through a hands-on consultation with a nurse or lactation consultant, and getting support at home. The efforts to breastfeed after a C-section are worth it, but it will require extra work.

BEST BETS FOR BREASTFEEDING

Below are a series of recommendations that will increase the chances of success in breastfeeding after a C-section.

1. Choose a hospital that is supportive of breastfeeding right after a C-section. In advance of your procedure, communicate with the nurses and doctors about your decision to breastfeed during recovery. This will require assistance, as you may not be able to handle the baby safely.

2. If the mother and child must be separated during the first few hours after birth, request a hospital breast pump to stimulate milk supply. The collected colostrum can be given to the baby.

3. Recruit help with home chores after the surgery, whether it's friends, family, paid help, or a postpartum doula.

4. Breastfeed often, as frequently as every two hours, to increase milk supply and prevent breast engorgement, which can lead to mastitis.

5. Breastfeed lying down, holding the baby on your side, to avoid added pressure and discomfort on the recovering abdominal area.

6. Try to avoid giving bottles for the first few weeks to prevent nipple confusion.

7. Be determined to stick to breastfeeding and don't hesitate to seek help from a lactation professional if it's not working.

6: Solid Foods:
A Growing Diet for Microbes

New Food Means New Microbes to Eat It

The early days and weeks of babyhood seem to go by so slowly, yet they also go by so fast. In the blink of an eye, the child who ate and slept all day begins to stay awake for longer periods of times, enjoying his surroundings and delighting mommy and daddy with cooing sounds and gummy smiles. Soon after, that same smile becomes very drooly, leaving a trail of slobber on every clothing item, bag, shoe, and piece of furniture you own. It's around this time that he starts staring at you while you're eating. After months of happily drinking only milk, he starts scrutinizing that weird, solid stuff that his parents put in their mouths, until one day he reaches for it, opening his mouth to get a bite of whatever it is that suddenly smells so good. This usually happens somewhere between four and six months of age, the time when solids foods should be introduced. Before this, babies obtain all the calories and nourishment they need from milk (including formula). Plus, as we discussed in the previous chapter, the intestine needs to mature for quite some time before it's ready for solid food.

The first time a baby eats solid foods marks an event that is usually photographed, recorded in baby diaries, and, nowadays, also shared on social media for the entire world to see. Who doesn't want a picture of their baby with goop all over his face, chair, hands, hair, and everywhere but his mouth? It's one of those wonderful "firsts" that no one wants to miss. Another event that fortunately *does not* get immortalized through pictures or diary entries is what happens in the hours or days after introducing babies to solid foods—those new changes in diaper content. It takes only one or two diaper changes to realize that things in your baby's tummy are changing and that diaper duty just got a lot more unpleasant.

Nothing influences the type of microbiota we harbor more than diet, so the introduction of solid foods marks a big change in the microbial community that calls a baby's gut home. With those first few spoonfuls of solid foods, the bacteria that specialize in digesting milk start being replaced by other species, and within a few months the infant microbiota begins resembling the microbial communities found in adults (and smelling like it, too!). The ecological changes in the gut microbiota after the introduction of solid foods vary greatly between individuals, but one common feature is an increase in microbial diversity.

The Boon of Diversity

Ecological diversity, also referred to as biodiversity, is determined by the number of different species in a particular habitat or place. In this case, introducing solid foods kick-starts a process that allows several new microbial species to feast on the new foods. Research from our lab and others has shown that by the time a child is one year of age, she will have approximately 60 percent more bacterial species in her gut than she did just seven months earlier—a huge jump in biodiversity. These new microbes have the ability to metabolize more complex sources of nutrients, as shown in a recent study led by Dr. Fredrik Bäckhed, a scientist at the University of Gothenburg in Sweden.

Just like many other ecosystems, microbial diversity in the intestine has been repeatedly proven to be a marker of good health. The same is true for a lake, a forest, an ocean, or other natural habitat—low biodiversity can be sign of an unhealthy and unstable ecosystem. Our intestine is no different; diseases like obesity, type 2 diabetes, and gastrointestinal disorders all share low microbial diversity as a common feature. Although more research is needed to clarify the extent to which microbial diversity promotes health, the message is clear: the typical Western lifestyle does not promote diversity.

The so-called Western diet—high in fats, sugars, and highly refined grains—is very strongly associated with a number of human diseases, and also with a less diverse microbiota. The reason for this low diversity is likely because most of the food eaten in these societies (and increasingly in developing countries) comes from very few species of plants and animals. Seventy-five percent of the world's food comes from only twelve plant species and five animal species. Amazingly, just three species—rice, corn, and wheat—account for 60 percent of the calories that humans obtain from plants, a shocking and sudden change in human practices when one considers that in the early 1900s there was 75 percent more genetic diversity in plants used for crops.

These days, everyone is eating the same stuff, and a lot of it, except in regions where a lack of economic development has kept people's farming and dietary practices more similar to those of a century ago. The effect this has on the gut was shown in a study comparing the microbiota of children living a rural lifestyle in Burkina Faso in West Africa to the microbiota of urban, city-dwelling children in Italy. The African children ate a high-fiber diet of vegetables, grains, and legumes, with an absence of processed foods, whereas the diet of the European children was saturated in sugars, animal fats, and refined grains, which also have more calories. The microbiota composition of the children from Burkina Faso was very different from—and much more diverse than—that of the Italian kids. Now, it's hard to argue that children from Burkina Faso have a healthier lifestyle than Italian children—they're more likely to suffer severe infections and malnutrition, and unfortunately have a lower life expectancy than a child born in Western Europe. However, they also have a decreased risk of suffering from the immune diseases that are becoming almost epidemic in the Western world. Coincidence? Not according to a mounting body of evidence suggesting that the early microbiota plays a very important role in the development of these disorders.

In an ideal world, children would harbor a rich and diverse community of microbes without the threat of severe infectious diseases, yet our current societal practices only address half of this equation (decreased infections). Given how well bacteria respond to diet, eating a variety of foods is most likely the best way to increase microbial diversity. Furthermore, there's no better time to establish a diverse microbiota than during the first 2–3 years of life. Remember, our microbial communities remain almost unchanged after early childhood, making it the optimal time to promote a diverse microbiota through diet. For example, don't feed a baby only rice cereal for weeks until the package is finished—offer them a variety of grains, including oats, rice, barley, quinoa, etc. It's also important to offer whole grains instead of refined ones. The Western diet is extremely low in fiber, and refined grains contain very little of it. Protein-rich legumes, such as lentils, beans, and peas, have an abundance of fiber and can be easily mashed for babies. Add more fiber by including vegetables in all or most meals and by offering nontraditional starchy vegetables such as sweet potatoes, parsnips, or cassava (tapioca), rather than just sticking to low-fiber veggies such as potatoes.

Understandably, the majority of people in developed societies won't crave these foods the same way they crave the texture of macaroni and cheese or the like, but the infant stage is the best time to introduce good dietary practices. For kids, eating healthy foods becomes a habit the same way cleaning their room does: by doing it frequently. When considering how to feed your growing baby and toddler, think not only about her, but also about her microbiota. Feed *them* a wholesome, varied diet rich in fiber and low in fats and sugar; both your child and her microbiota will thank you for it.

When, What, and How Much?

Knowing when to introduce solid foods and what foods to first offer your child can be confusing because the rules keep changing. For example, Claire experienced a change in guidelines between her two children, who are not even two years apart. With her daughter, Marisol, she was instructed to introduce solid foods at six months of age, starting with cereals and slowly moving into vegetables, fruits, and meats. Just over a year and a half later, with her second child, their family doctor recommended that she give him solids between 4–6 months, upon him showing signs of readiness. Then last year, a friend of Claire's had a baby and the guidelines had changed again, with the advice to start at 4–6 months and give meats first. Surely, certain things still need to be figured out when it comes to the timing of solid food introduction, although one thing is clear: there's wiggle room for starting at the four-, five-, or six-month mark, depending on when the baby shows interest and readiness.

Before 4–6 months of age, babies don't need any other nourishment than milk, but approaching six months of age their levels of iron start to decrease. Iron is an incredibly important mineral—it carries oxygen from the lungs into the rest of the body, among several other important functions. Having low levels of iron in the blood is known as iron-deficiency anemia, a condition that can be prevented by offering iron-rich foods to babies at around six months. Iron is also important for brain development and an iron deficiency is associated with a lower IQ. Current recommendations are to offer foods such as meat, meat alternatives (eggs, tofu, or legumes), or iron-fortified cereal. However, getting the full daily requirement of iron would take half a cup of iron-enriched cereal, which is a huge amount for a baby just starting solids, and it's not as well absorbed as natural sources like meat, anyway (which is unfortunate news for vegetarians, since iron from nonanimal food sources is also not absorbed as easily as iron from meat). Note that babies given formula should receive 2.5 ounces of iron-fortified formula per pound of body weight each day, gradually decreasing this amount as the baby gets older and eats more solid food.

Although the eventual goal, as mentioned previously, is a diverse diet, there are compelling reasons to nurture a baby's palate slowly. Foods should be introduced in small quantities and, ideally, one at a time. Initially, the baby will have just a taste, then move up to a teaspoon of food, then a whole tablespoon, and so on. This gradual increase should follow the baby's cues: if he wants more, offer a bit more; if he turns away, hold off for now. Offering foods one a time allows the baby's digestive system to get used to one ingredient for a couple of days before it tries a different one. It also allows a parent to detect if one particular food does not sit well with the baby or if it causes an allergic reaction (more on food allergies in the next section).

Within a few weeks, a baby will have tasted lots of foods that can start to be mixed and matched. This is when parents should think about variety as the central theme of their child's diet, offering options for every food group (meats, grains, fruits, and vegetables) and choosing foods rich in fiber as much as possible. If a baby doesn't like a food, don't force it, but definitely try again later. Some babies may need to be given a new food as many as 10–15 times before they'll actually eat it. A quick look at what babies around the world eat is proof that babies do eat whatever they're offered, even if it needs to be offered many times (see First Foods Around the World, page 100).

Another important thing to keep in mind is that the introduction of solid foods doesn't mean that babies should be weaned from milk. On the contrary, it has been proven beneficial to continue to breastfeed on demand until the baby is two years old, or even longer if the mother and child still want to. A recent study from the laboratory of Dr. Fredrik Bäckhed showed that the microbiota of babies who are weaned early shifted more quickly to an adultlike microbiota than babies that continued to breastfeed. Given that the presence of milk-loving bacteria has such strong and beneficial immune effects, it's thought that delaying the maturation of the microbiota is beneficial to young children. As such, it's recommended to continue to breastfeed or to provide a formula with probiotics to babies eating solid food.

Another way to boost your baby's levels of probiotic species is to introduce them to fermented foods such as yogurt or, even better, kefir. Kefir is a drink very similar to yogurt, except it has a more watery texture and is slightly more sour (and has a ton of probiotics in it; its advantage over yogurt is that it has a lot more different species). For either yogurt or kefir, it's optimal to choose a brand low in sugar and without artificial sweeteners. The fewer refined sugars your baby drinks the better, both for the baby and for the trillions of microbes feasting on everything that ends up in that tummy.

Dangerous Eats

Except for poisonous foods and choking hazards, food is safe, right? Ideally, yes . . . but it turns out that certain foods agitate the immune system and elicit an allergic reaction in some people. Many foods are known allergens—the usual suspects are wheat, eggs, milk, peanuts, tree-nuts, fish, shellfish, strawberries, sesame, and soy. These allergic reactions tend to run in families, so if a parent or a sibling has a specific food allergy, there's an increased risk that the new baby will develop one, too. However, the genetic basis for food allergies does not explain the huge increase in cases in the past generation. According to a study by the CDC, food allergies in children increased 50 percent between 1997 and 2011, an enormous jump in incidence that has one in every twenty kids suffering from a food allergy. Twenty years ago it would have made no sense to see a note on a package of gummy bears warning that it may contain traces of peanuts, whereas nowadays millions of parents have to scrutinize safety signs on packaged food, because without one, it can't go in the lunch box.

Food allergies have a large spectrum of symptoms, from mild ones such as an itchy mouth, to severe ones like anaphylaxis—a potentially deadly reaction. Few things are scarier than imagining your child suffering a serious allergic reaction, yet a child visits an emergency room for this reason every three minutes in the US. Malcolm and Jeannie are the proud parents of three teenage boys, and one of them, fourteen-year-old Callum, has a severe allergy to certain nuts. Jeannie found this out when Callum was about fifteen months old, after she offered him a cashew. Callum quickly became fussy and would not stop crying until he fell asleep. A few minutes later he woke up irritated and they noticed he was covered in hives. They phoned his pediatrician right away, who urged them to call 911 and wait for an ambulance. Once the first responders arrived, they assessed Callum and took him to the ER, where they treated him and gave him a referral to an allergist. A few weeks later, through a series of tests, they found out that Callum was not only allergic to cashews, but also to pecans, almonds, and peanuts. Malcolm and Jeannie were given the same (and only) advice that thousands of parents receive after their child is diagnosed with a food allergy: avoid the problem food completely and carry an EpiPen in case accidental exposure occurs.

It's not clear why there's been such a jump in the number of food allergy cases, but it's likely that exposure (or lack thereof) to early-life microbes plays a role. For example, babies born and raised on farms have a lower chance of developing food allergies, and breastfeeding for the first six months reduces the risk and severity of certain food allergies. In addition, epidemiological studies are showing that the time at which these foods are introduced is very important. In Israel, where peanut allergies are ten times less prevalent than they are in the UK Jewish population, babies are given a popular healthy snack called Bamba as soon as they can safely eat it. Nothing seems out of the ordinary until one realizes that 50 percent of the ingredients in this snack are peanuts. As any parent that has raised a child in the past twenty years can confirm, giving peanuts to a baby under one year of age is a huge no-no. Or is it? According to the latest research, in trying to protect our children from food allergies, it appears that we may have been doing the opposite.

If navigating the recommendations about when to start solid foods is confusing, deciding on when to introduce foods that are known to induce allergies is even harder. It's also hard for pediatricians and other health practitioners to keep up with the current lines of thought and offer the right advice (sometimes they have to offer different advice to the same parent for each individual child!).

When food allergies started becoming more common around the 1990s, experts agreed that delaying the introduction of these foods would reduce the likelihood of developing an allergy. In the year 2000, the American Academy of Pediatrics issued guidelines that infants should wait until age one to have cow's milk, until age two to have eggs, and until age three to have shellfish, fish, peanuts, and tree nuts. Eight years later, the guidelines were revised and the accompanying statement cited little evidence that delaying the introduction of these foods was beneficial in preventing food allergies, yet it didn't include new recommendations, leaving parents and doctors in limbo. "As these guidelines were implemented we've seen a paradoxical increase in foods allergies in young children, especially with peanut allergies," said Dr. Anna Nowak-Węgrzyn, a professor of pediatrics at the Icahn School of Medicine at Mount Sinai in New

York. It's taken a few years to gather enough evidence to show that delaying the introduction of these foods is not only ineffective, but it may be making things worse.

A recent study published in the *New England Journal of Medicine*, one of the most prestigious medical science journals, revealed that children who received a delayed introduction to peanuts had an increased risk of developing peanut allergy, compared to children that encounter them early. Just how early? Very—between four and seven months of age. Because of this landmark study, as well as others, the new recommendations issued by the American Academy of Asthma, Allergy and Immunology (AAAAI), Canadian Paediatric Society (CPS), and Canadian Society of Allergy and Clinical Immunology (CSACI) state that allergenic foods should be introduced in the same way as other foods: slowly and gradually, starting at 4–6 months. The AAAAI recommends that once an infant has been given a few nonallergenic foods (meat, vegetables, etc.) the allergenic ones can be offered without delay, ideally before seven months of age.

These new guidelines not only follow what the evidence says, but they also make sense based on what's currently known regarding the immune system of young infants. During the first months of life, exposure of foods to the gastrointestinal immune system encourages immune tolerance. In addition, once an allergenic food is introduced, it appears to be equally important to maintain frequent, regular ingestion early on in order to maintain tolerance and truly prevent a food allergy.

It's important to take into account that one of the factors keeping the infant's immune system in this tolerant state is the presence of a bunch of microbes that specifically work to promote immune tolerance. Thus, breastfeeding throughout this stage or supplementing formula with infant probiotics will help maintain higher levels of these hardworking microbes. The same guidelines issued by the AAAAI recommend breastfeeding at least until four months of age, although it could be argued that, based on the current understanding of how allergies develop, extending breastfeeding beyond four months is probably beneficial, too. Another concept that has changed with these guidelines is the restriction on pregnant mothers eating allergenic foods. New evidence shows that such a restriction does not prevent the development of allergies in babies.

Dos and Don'ts

◆ **Do—** look for signs of readiness for your baby to start eating solids: he can sit up without support; he doesn't automatically push solids out of his mouth with his tongue; and he seems interested in food and watching *you* eat. Try offering solid foods between 4–6 months of age once you notice your baby is ready.

◆ **Do—** start slowly, always following your baby's cues, and increasing amounts gradually. Begin with one food at a time, trying the same food for 2–3 days. This will both help your baby's digestive system get used to a particular food and help detect a possible food allergy.

◆ **Do—** diversify your baby's microbiota. After single-course solids have been eaten for a few weeks, try varied choices from all food groups. For meats, offer different meat sources as well as meat alternatives such as eggs, tofu, or legumes (beans, peas, and lentils). For grains, choose wheat, rice, oats, corn, barley, rye, quinoa, etc. Pick whole grains and whole grain flours, as they add substantial amounts of fiber. Provide various types of vegetables and fruits in every meal and consider serving nontraditional starchy vegetables, such as sweet potatoes, parsnips, or cassava.

◆ **Do—** keep sugary foods to a minimum, especially juice. A baby with a sweet tooth will likely become a toddler with a sweet tooth.

◆ **Don't—** stop breastfeeding once your baby starts eating solids (if possible). A baby continues to benefit from breast milk until she's two, so the longer you're able to breastfeed, the better. If formula feeding, look for a brand that contains probiotics, or administer them separately in the form of pediatric probiotic drops. This will help delay the process of your baby's microbiome becoming adultlike too early.

◆ **Don't—** delay the introduction of allergenic foods. Offer peanuts, soy, shellfish, etc., after less allergenic foods have been tolerated, between 4–7 months of age. Do this slowly and using the "one at a time" rule, just like with any other food.

◆ **Do—** keep up with the most current medical information regarding solid food introduction and ways to prevent food allergies. The guidelines will likely continue to change as more studies get published and revised by medical associations.

FIRST FOODS AROUND THE WORLD

For almost half a century the first food staple for North American babies was processed single-grain cereal. Fortunately, the recommendations have changed and babies are now encouraged to eat veggies, meats (or meat alternatives), and iron-fortified cereals. A look at the foods that babies around the world taste for the first time makes rice cereal seem terribly boring—and it proves that babies will eat anything.

Chinese babies get lots of rice in their first bites, but it's often mixed with tofu, fish, and vegetables. Japanese babies start on solids similar to Chinese babies, but also enjoy seaweed. Most Japanese babies have had raw fish before they turn two! Filipino babies often begin solids with rice simmered in broth, chicken bits, onions, and garlic. East Indian babies start eating lentils and rice, spiced with coriander, mint, cinnamon, and turmeric as early as six months of age. Mexican and Central American children munch on corn tortillas, mashed beans, and vegetables as first foods. In Mexico, babies eat chilies before they turn one, and candy is often mixed or sprinkled with chilies, lemon, and salt.

So the next time your child refuses to eat something you offer, remind her that she should be thankful she's not in Tibet, where babies eat barley flour mixed with yak butter tea!

7: Antibiotics: Carpet Bombing the Microbiota

The Antibiotic Paradox

Pam was excited about her upcoming delivery. The pregnancy had gone well, but then, when the labor finally came on, things began to go wrong. Pam was fully dilated, but when she started to push, the baby wasn't coming down the birth canal properly. The obstetrician tried forceps, but the baby was really stuck. Pam then underwent a C-section, followed by antibiotics, just to be safe, to prevent any infections that might result from the procedure.

Pam is one of the few women who experience a C-section on top of full length of labor, a huge ordeal. Unfortunately, Pam's troubles weren't over yet. After the difficult birth, she returned home from the hospital with her beautiful newborn daughter, but then had trouble breastfeeding. Her nipples were cracked and she developed mastitis, followed by diarrhea. The diarrhea became so severe that Pam had to be hospitalized for dehydration and high fever. She tested positive for *Clostridium difficile* (also known as *C. diff*), a severe intestinal infection that occasionally follows antibiotic treatment, which landed her in isolation at the hospital (to help prevent the spread of *C. diff*). She was given another antibiotic (vancomycin) to treat the *C. diff,* and eventually things began to return to normal.

Pam's story highlights both the pros and cons of antibiotics. They're great drugs for controlling microbial infections, but we now realize they can also cause problems and side effects that we didn't fully appreciate before, which is making us rethink their applications.

Wonder Drugs of the Twentieth Century

Arguably, antibiotics have had the greatest effect on improving human health and longevity of any class of drugs used in the twentieth century. Their invention was truly magical, easily treating diseases that previously could lead to death. Before antibiotics, 90 percent of children with bacterial pneumonia would die. Children with strep throat were placed on bed rest to try to avoid the dreaded complication of rheumatic fever. Talk to your grandparents—or anyone who grew up in the pre-antibiotic era (prior to 1945)—and you'll realize how scary a simple infection could be.

The word *antibiotic* comes from Greek, meaning "against life," and we use the term to describe drugs that work against microbial life. This includes chemicals called antibacterials, antivirals, antiparasitics, and antifungals. However, we tend to use antibiotic and antibacterial interchangeably, and antibiotics usually refer to drugs that target bacteria as opposed to other microbes such as viruses or fungi. Some antibiotics are broad spectrum (meaning they target many types of bacteria; see Going Bananas with Amoxicillin, page 115); others are narrow spectrum (they target fewer microbes, but still don't specifically target a single microbial species).

The discovery of antibiotics dates back to the early 1900s. A chemist named Paul Ehrlich had the idea of a "magic bullet" that could target a disease-causing microbe without harming the human host. He tenaciously tested many compounds and synthetic dyes, screening them to find a compound that would target the bacterium that caused syphilis. After Herculean efforts, he discovered a compound that could be used to treat this infection. Although the drug did have serious side effects, such as rashes, liver damage, and "risk of life and limb," the cure was better than the disease. Ehrlich's work laid the foundation for the discovery process that has given us so many new antibiotics.

By far the most famous antibiotic discovery was made by the British scientist, Sir Alexander Fleming. Apparently, Fleming wasn't the most organized of scientists and in 1928 he came back to his lab after a vacation to find lots of petri dishes that he had left out covered with bacteria. He began to sort through all this mess when he noticed something out of the ordinary in a petri dish that contained *Staphylococcus aureus* (the bacterium that causes skin boils and abscesses). One area on the petri dish was free of bacteria, but it had a blob of mold instead. This mold, which was later identified as the fungus *Penicillium*, produced a substance that inhibited growth of the bacterial pathogen. Penicillin is a complex molecule, and it took almost two decades for chemists to figure out how to synthesize it. With the outbreak of World War II, there was a huge need to control bacterial infections from the battlefield. In the pre-antibiotic era, wars were terrible for infections (in the American Civil War, more people died from infections than bullets). So during World War II, with infections raging, penicillin was invaluable. Initially, it was so precious that it was kept for military use only, and it was recovered and re-isolated from patients' urine in order to reuse it.

However, penicillin production was rapidly scaled up, and by 1945 it was mass-produced and distributed to the general population, saving countless lives.

This breakthrough changed the world of infectious diseases and how they were treated, opening up therapies for diseases that were previously considered untreatable and often fatal. Several new classes of antibiotics were discovered in the 1950s–1970s, mainly from soil organisms. Soil organisms produce antimicrobials to kill their neighbors since they're in competition for scarce nutrients. Microbiologists collected soil samples from the far corners of the world and tested them for antimicrobial activity. When promising compounds were found, chemists modified and tweaked them to enhance their activity or uptake into microbes. These semisynthetic compounds formed the backbone of the antibiotic industry, providing many potent new antimicrobial drugs that could be used to treat a variety of infectious diseases.

Like all good things, we just couldn't get enough. There are at least 150 million antibiotic prescriptions written in the US each year (that's one for every two people.) In 2010, the estimated global consumption of antibiotics was 63,000 tons, which will increase by 67 percent in 2030 to over 100,000 tons. Much of this is due to their increased use as growth supplements in livestock (80 percent of the US's antimicrobial usage is in livestock). It turns out that low doses of antibiotics enhance the weight gain of farm animals. Europe has stopped this practice, but unfortunately Canada and the US have not, and developing countries are just beginning to use antibiotics in livestock management. Some of the consequences of this are discussed in chapter 10, where we look at how this practice affects childhood obesity.

So what effect does dumping into the world thousands of tons of chemicals that kill microbes have on the microbiota? As we shall see, the microbes have responded, and it's casting serious shadows on these wonder drugs.

Resistance Is Futile

Well, maybe not if you're a microbe. As mentioned previously, most antibiotics come from compounds made by soil microbes to give them a competitive advantage. However, microbes must have a way to resist killing themselves with these toxic molecules. As a result, microbes have developed resistance mechanisms to accompany antibiotic production. We call this antimicrobial resistance, and microbes have been doing this for as long as they've been producing antibiotics. For example, antimicrobial-resistant genes have been found in human remains that were frozen in permafrost, obviously millennia before antibiotics were discovered. Resistance genes have also been found in environments where there has been no human contact, such as underground lakes below sheets of ice. As soon as penicillin was discovered, it was also realized that microbes exist that can resist its killing effects.

The other thing we need to remember about microbes is that, unlike us, they regularly exchange DNA with one another, allowing them to rapidly adapt and evolve during their lifetime (which can be as fast as twenty minutes, but is often a few short hours). Think of the microbes in your gut as being hooked up to a genetic Internet—they exchange genes much like we download songs and apps. Having an "antimicrobial resistance app" would be a lifesaving feature if you were getting hit over the head with a lethal antibiotic. In response to our massive use of antibiotics, microbes have spread resistance genes like wildfire to other microbes (their "must-have" apps!), with antibiotics providing a strong selection pressure to get that app (live or die). We now see massive microbial resistance to all major antimicrobials that are used extensively, with resistance arising within a year or two, often making the drug obsolete within 3–5 years.

We are also discovering more and more microbes that are resistant to most, if not all, antibiotics—we call them "superbugs." They cause infections that we could previously treat, and they're wreaking havoc in our hospitals and health care systems worldwide. They go by lovely acronyms such as MRSA, XDR TB, MDR E. coli, and VRE, and are causing a major rethink in how we use antibiotics.

Since 2009 only two new antibiotics have been approved, and the number in the pipeline continues to shrink. Most pharmaceutical companies have either drastically downsized or closed their antibiotic discovery divisions. This is creating the perfect storm: no new drugs, and the ones we have no longer work. The World Health Organization recently summed it up nicely, saying that antimicrobial resistance is a "serious threat [that] is no longer a prediction for the future, it is happening right now in every region of the world and has the potential to affect anyone, of any age, in any country." These strong words suggest that we're headed for a post-antibiotic world, taking us back to the fears of mortal infections common to our great-grandparents.

"Mommy, My Ear Hurts!"

These words strike fear into any parent's heart, knowing it usually means a sleepless night, if not a trip to the ER in pajamas. Ear infections, called otitis media, are quite common in young children, and are usually treated with antibiotics. However, it isn't always clear that antibiotics are warranted, which is confusing to parents (and physicians). Take the story of Jack, a two-year-old who had developmental difficulties and was also prone to recurrent ear infections. After a particularly sleepless night, Jack's mother was convinced her son had yet another ear infection and took him kicking and screaming to see his pediatrician. Diagnosing otitis media usually means visually observing the eardrum (called the tympanic membrane) to see whether the eardrum is bulging, which indicates fluid in the middle ear, and whether it looks red, which suggests inflammation. Redness of the eardrum can also result from a child crying and does not always mean an infection is present. The pediatrician had difficulty assessing Jack's eardrums, as he was a combative child, and he also had wax in his ears. She thought the ear looked red, and considering the mother's description of the symptoms, prescribed a common antibiotic used to treat ear infections. Unfortunately, this resulted in Jack subsequently getting *C. diff* diarrhea, which required him to go on metronidazole (a strong antibiotic also called by the brand name Flagyl). The first course of metronidazole did not work. The second time, his mother also wisely gave him a yeast probiotic (see Probiotics with Antibiotics—an Oxymoron?, page 112), and he finally got better. After that, an ear, nose, and throat surgeon placed drainage tubes in Jack's eardrums in order to decrease the number of ear infections, rather than use multiple courses of antibiotics to treat them.

Treating otitis media with antibiotics is controversial. First of all, it's often overdiagnosed. The eardrum can be red from crying or due to a viral infection, so antibiotics wouldn't work in those instances. Anyway, most cases of ear infections caused by bacteria resolve on their own. Studies in the Netherlands indicate that seven kids with otitis media would have to be treated with antibiotics in order for one to benefit from antibiotic therapy. There is a very small risk that otitis media, if left untreated, will progress to more serious illnesses such as mastoiditis, a dangerous infection of a bone that sits behind the ear. However, this is uncommon. The Dutch study calculated that only 1 in 2,500 children develop a complication from otitis media; if they were all treated, it would result in a lot of kids receiving unnecessary antibiotics. Seventy-five percent of children have at least one episode of acute otitis media before their first birthday, so this is certainly an infection every parent will encounter at some point. It's usually preceded by an upper respiratory infection (more often than not a viral cold), which plugs up the Eustachian tubes that drain the middle ear, which then fills up with fluids, making a perfect broth in which bacteria happily grow.

Most pediatric organizations have developed guidelines for treating otitis media, along with adopting a cautious use of antibiotics. They generally suggest a "watch and wait approach," especially if the child is older than six months, is otherwise healthy, and has mild symptoms (no major fever, etc.). Doctors will also ensure that parents have access to painkillers to help the child ride things out. Generally, the recommendation is to wait 48–72 hours, and then follow up if the infection has not resolved. This results in approximately only one-third of children getting antibiotics, which is much more reasonable than administering them to every single child who shows early symptoms.

The other current practice is how pediatricians aim to "hit hard and fast" if they're going to use antibiotics. Five days of antimicrobial treatment are at least as effective as ten days of antibiotics in children older than two years of age. A longer span of treatment may be needed in complicated cases, but the general idea is to expose your child to fewer antibiotics, if medically reasonable.

Breastfeeding has been shown to decrease ear infections, probably because the maternal antibodies in breast milk help protect against infections. Bottle-feeding causes a child to suck hard on the negative pressure inside the bottle, which then causes negative pressure inside the ear, which may draw fluid and microbes into it (fully ventilated bottles can avoid this problem). Similarly, extensive use of a pacifier in children younger than three years old increases the risk of otitis media by 25 percent, presumably for the same reasons. Since ear infections usually follow respiratory infections, minimizing the time young children (particularly those younger than twelve months) spend in day care, and thus minimizing their inevitable exposure to colds, also decreases the risk of ear infections. Infants are routinely given a pneumonia vaccine (Prevnar), which may also help make a child less susceptible to ear infections. Influenza vaccination can help as well. Maternal smoking in the first year of life is a significant risk factor, too, especially in children that had a low birth weight.

Wonder Drugs That Aren't So Wonderful

In the title of this chapter we used the expression "carpet bombing," which is a military term that entails bombing a defined area with indiscriminate destruction and lots of collateral damage, in hopes of destroying the desired military targets. Unfortunately, this description can also be applied to antibiotics and their effects on the microbiome. We now know that antibiotics cause massive disruption to the microbiota (naturally, as they're designed to kill as many microbes as possible). This indiscriminate killing of many bystander microbes, in addition to the desired infectious agent, has unintended consequences. We're also starting to realize that we may be wiping out microbes from our society before we even realize that they're beneficial. As detailed in Dr. Martin Blaser's book *Missing Microbes*, previous generations had much more diverse microbiota, and our quest to kill them all may have serious consequences for future generations. Who would have thought that we might have to put microbes on the endangered species list?

As discussed more in chapter 13, multiple courses of antibiotics in the first year of life lead to a significant increase in asthma, as they affect the microbes involved in ensuring a healthy maturation of the immune system. Similarly, in chapters 10 and 11 we discuss how antibiotics increase the rate of obesity and subsequent type 2 diabetes by altering the microbiota involved in nutrient uptake and weight gain. For these diseases, there's a direct correlation with the number of courses of antibiotics, with up to a 37 percent increased risk with multiple uses of certain antibiotics.

Unfortunately, even short-term pulses of antibiotics during early life can have an effect on the microbiota. Experiments in lab animals have shown how these pulses can cause developmental changes in mouse pups, with increased weight gain and bone growth. In studies of children, the effects of antibiotics on the microbiota are still observed six months later, and repeated use of antibiotics shift the microbiota further and further away from its initial composition. What's more, the presence of antimicrobial resistance genes can still be detected 1–3 years after the antibiotics are stopped.

As we saw in the two examples mentioned earlier, there's a significant risk of getting *C. diff* infections following antibiotics, due to the removal of microbes that normally prevent *C. diff* from taking hold in the gut. This is discussed in chapter 16 in more detail, along with the exciting (and somewhat off-putting) concept that fecal transplants can successfully treat such infections. Our lab has shown that antibiotics also make mice more susceptible to diarrheal infections caused by pathogenic *E. coli* and *Salmonella*. This is called "competitive exclusion," with the concept that the good bugs make it much harder for the bad ones to take hold in the very crowded intestinal world. Our lab also showed that antibiotics cause a thinning of the mucus layer, which serves as a protective barrier against infections and other inflammatory intestinal disorders. This is presumably because the antibiotics affect microbiota that normally eat mucus for an energy source. Another example of competitive exclusion can be seen in urinary tract infections, which often arise in women following a course of antibiotics for a different infection; the protective vaginal microbiota is disrupted, thereby allowing a pathogen to infect the urinary tract.

It turns out that antibiotics have other effects that we previously never even dreamed about. In some studies conducted in our lab, we treated mice with standard antibiotics and measured how this affected small molecules (called metabolites) in the feces and liver. Remarkably, we found that this treatment significantly changed about 60 percent of the mouse's metabolites. These are molecules involved in normal body function, including hormones, steroids, and other chemicals that allow our cells to communicate with one another. Needless to say, we were shocked when we realized that antibiotics could impact our normal physiology that much.

Immune cells are also innocent bystanders, affected when antibiotics cause changes to the microbiome. For example, in an animal model, it was shown that neutrophils, which are key microbe-hungry immune cells that clear away pathogens, were decreased in newborns treated with standard antibiotics. The researchers found that with antibiotic treatment, certain classes of microbiota that stimulate the immune system were removed, thereby resulting in a decrease of neutrophil production.

Another common side effect is diarrhea, which plagues up to

one-third of people following a course of antibiotics. Even in developing countries where diarrhea is much more common, higher levels were detected in young children (six months to three years old) following antibiotic treatment, although children who were exclusively breastfed for the first six months were protected from this side effect.

Physicians now realize the issues associated with antimicrobial resistance and are beginning to take into account the detrimental effects of antibiotics on the microbiome. Pediatric associations have developed guidelines for "antibiotic stewardship," optimizing the use of antibiotics while minimizing the unintended consequences. This certainly helps with antibiotic abuse, and it also explains why your pediatrician may be more reluctant to reach for the prescription pad than in previous years. Your doctor may ask you to wait until she gets the results back from a strep throat swab before starting antibiotic treatment—she's just being prudent, despite your screaming kid's input on the decision.

Ironically, many parents are aware of this antibiotic dilemma, and will question physicians on whether an antibiotic is necessary. One option is to get the prescription, but wait and see if the infection improves on its own before using the antibiotic.

Probiotics with Antibiotics— an Oxymoron?

Antibiotics kill microbes, so why would one intentionally take probiotics (which are live microbes) while taking antibiotics? In certain cases it's actually a good idea. As discussed previously, diarrhea and *C. diff* are common complications of antibiotic use (*C. diff* causes about a third of the types of diarrhea). A recent study showed that if certain probiotics (although not yogurt) were taken along with anti-biotics, 42 percent of the participants were less likely to have diarrhea. In previous years, the counsel was to wait until the antibiotics were stopped before taking probiotics. However, *Lactobacilli* probiotics are now being marketed specifically to be taken along with antibiotics, although you're still encouraged to wait 1–2 hours after ingesting the antibiotic before taking the probiotic. An interesting concept is to take a yeast probiotic when taking antibacterials, since antibiotics do not kill yeast, as they are not bacteria. In a major meta-analysis (a summary of publications) of thirty-one probiotic studies, it was concluded that *Saccharomyces boulardii* (a yeast) worked well to decrease antimicrobial-associated diarrhea *and* to decrease *C. diff* infections. The same study found that the bacterial probiotic *Lactobacillus rhamnosus* GG helped prevent diarrhea in children, but had no effect on *C. diff* infections. They also found that mixes of two probiotics could have some efficacy against antibiotic-associated diarrhea.

The other compelling conclusion is that one needs to take large doses (more than 10 billion live microbes a day) to have any effect. Since the onset of antibiotic-induced diarrhea is usually 2–8 weeks after taking antibiotics, most patients are encouraged to continue using probiotics even after finishing the antibiotics. Collectively, this work suggests that certain probiotics can help prevent diarrhea when taken with antibiotics, given the caveats above.

Antibiotics have gone from being miracle drugs that could bring a dying person back to life to being used indiscriminately for every fever in a child. We now live in an era where their intense use has backfired in the form of antibiotic resistance. On top of that, few things shift the developing microbiome of an infant or child more than antibiotics, potentially affecting their immune system permanently. Fifty years ago no one saw any of this coming, but we're facing a reality where an antibiotic must be seen as a drug of last resort that often requires restoration of the microbiota following its use.

Dos and Don'ts

◆ **Don't—** assume that all infections have to be treated with antibiotics. Upper respiratory tract infections and colds are often caused by viruses, so antibacterials won't cure them. Most sore throats, especially if the child also has a runny nose and cough, are caused by viruses and don't need antibiotic therapy. If your child has a mild ear infection, it's reasonable to watch and wait for a few days to see if it gets better on its own before starting antibiotic therapy.

◆ **Do—** consider giving probiotics to your child if he is being given antibiotics. These could include *Saccharomyces boulardii*, a yeast-based product that decreases *C. diff*, or *Lactobacillus rhamnosus* GG, or a mixture of probiotics. This assumes the child is not immunocompromised nor has any other underlying conditions.

◆ **Do—** be a thoughtful steward regarding antibiotic usage. Discuss with your pediatrician why she is suggesting an antibiotic (she presumably has her reasons). Antibiotics are a remarkable treasure to medicine, but their abuse is really denting their magic, and we now realize they aren't without their own detrimental effects.

◆ **Do—** make your child less susceptible to ear infections by using a ventilated bottle for milk and avoiding excessive use of pacifiers in children under the age of three.

GOING BANANAS WITH AMOXICILLIN

The most commonly used children's antibiotic is amoxicillin, a type of penicillin that targets the walls that surround bacterial cells. It's often formulated to taste like bananas, with the idea that kids will like it. However, children still tend to turn up their noses at the taste, and it can be a real struggle to get them to take it! One of the most common and very appropriate uses of amoxicillin is to treat ear infections that haven't cleared up on their own after two days of watchful waiting. Recommended treatment is usually for 5–7 days.

Ear infections are commonly caused by several different bacteria, including *Streptococcus pneumoniae* (commonly called "pneumococcus"), *Haemophilus influenzae* (commonly called "H flu"), and *Moraxella catarrhalis*. However, amoxicillin kills not only these pathogens, but many other bacterial species, which has a major effect on the microbiome.

Unfortunately, amoxicillin is also commonly used, inappropriately, for viral upper respiratory infections such as the common cold and the flu. Because it targets a bacterial structure, it has no effect on viruses. And of course, resistance has become common—it used to be the first line of treatment for bacterial urinary tract infections, but now resistance is too common and other antibiotics must be used.

8: Pets:
A Microbe's Best Friend

Love at First Lick

When Nathan and Carol left the hospital with their brand-new baby, Rory, they were a bit nervous about how Milo, their three-year-old Labrador retriever, would react upon meeting the new addition to the family. On one hand, they were worried about not having enough time for Milo, who up until then would go with them on hours-long hikes and swims and on frequent camping trips. With a baby in the house, the dog was being moved to second place in the attention contest. On the other hand, Milo, like almost any other lab on the planet, liked to jump up, sniff, and lick everyone he met, and months of doggy school hadn't curbed his habits. Carol and Nathan were concerned about this seventy-five pounds of bouncing friendliness around wee little Rory. On top of that, they had conflicting opinions about how close Milo and Rory should be. Carol was happy to have him sniff and lick Rory, but Nathan, and especially Nathan's mom (a frequent visitor), were not. Nathan didn't grow up with pets in the house, so even though he was a committed owner to Milo, he would shoosh him off the couches, the nice living room carpet, and, most of all, the bed. Carol respected that Nathan was not as okay with dog slobber and hair as she was, but when Nathan wasn't home she would curl up with Milo on the couch to watch a movie, or invite Milo to bed if Nathan was spending the night away from home. Needless to say, they still needed to sort out the details on their new status as a family of three humans and a dog.

They decided to talk to a dog trainer who had helped a lot with Milo's behavior when he was a puppy. She said that although labs are usually great pets around kids, there were a few things they could do to make the transition easier and safer. First, she recommended that Milo meet Rory for the first time outside the house and have him leashed. She also said that the first encounters at home should be supervised and that they should continue giving attention to Milo when the baby was around. As for any recommendations regarding licking the baby, she quickly said, "You're on your own on that one! Everyone does it differently."

When it came time to introduce Milo to Rory, they arranged the meeting at a nearby park. Nathan's mom brought Milo on a leash, but upon seeing his owners, Milo broke free of her hold, running to greet his owners and jumping at them with excitement. Despite their efforts to calm him, when presented with his new "brother," Milo sniffed him from top to bottom, then touched Rory's face with his wet nose and very gently licked Rory's pink cheek. "No licking, Milo," said Nathan, pushing Milo's face away. "It's just a kiss, Nathan," said Carol, knowing at that moment that Rory and Milo were going to be best friends.

From the Wild to Our Couches

The partnership between humans and dogs goes way, way back. Well before humans settled down to farm, dogs would roam with packs of hunters and gatherers, possibly scavenging leftovers from their human companions' hunted game and other food. Dog fossil remains have been found in caves dating as far back as 16,000 years ago, at a time when humans would compete against saber-toothed cats to hunt mammoths across a frozen landscape. Archaeologists aren't in agreement on the exact location and timing of dog domestication, but one thing they all acknowledge is that dogs were the first species domesticated by humans, including all plants, insects, and other animals.

In the beginning they were merely tamer wolves, still feral and only seeking human interaction as a way to get food, but the two species quickly became very close. Through thousands of years of living with humans, dogs became reliable guards, hunters, herders, and carriers, and they developed an uncanny ability to communicate with humans, superior even to how chimpanzees (or any other ape) can pick up on human cues.

Cats, on the other hand, became domesticated after humans took to the farming lifestyle. They likely became useful to people as a way to control rodents in granaries, and probably acquiesced to the deal in exchange for food, shelter, and play. Compared to dogs, cats only recently split off from wild cats and some even breed with other wild feline species to this day. Their genome hasn't changed as much as the dog's genome has, and they still require a high-protein diet and can't digest human food very well. In that sense, cats remain only semidomesticated, despite having lived with humans for at least 9,000 years. Every now and then we see their wildness, whether it's when they turn up on the doormat with the offering of a dead mouse, bird, or lizard, or how they disappear for multiday escapades, only coming home to get fed. It's not surprising that cats, at least in most Western societies, are kept as full-time indoor pets, seldom allowed to roam and test out their wild sides, for fear that, one day, they simply might not come home.

Some people keep pets for their useful qualities, but most modern societies keep cats and dogs for companionship. They can require a bit of work, but their loyal friendship, silliness, and unconditional love usually make it worthwhile. There are many obvious ways that having a pet can improve your lifestyle; dogs, for example, promote exercise (daily walks, rain or shine!), encourage sociability ("Hi, what's your dog's name?"), or simply make you happy (nothing beats a wagging tail and smiling face every single time you walk through your door!). As if those aren't reasons enough, we're now beginning to learn that pets, especially dogs, also bring health into your life by bringing the outside indoors. Yes, all those dirty paws on floors, carpets, and furniture, all those stinky smells that won't go away are worth it—within all that dirt there are millions of microbes that make our clean lives that much closer to the outdoors.

The influence dogs have on our microbiota was recently documented in two studies, wherein scientists found that owning a dog (but not a cat) that roams around outside changes the composition and diversity of the human microbiota. One study showed that the microbiota among family members is more similar in families that own a dog compared to dogless families. The same study also found that the skin microbiota of dog owners had bacterial species also found in dog mouths and in the soil. The similarities between dogs and their owners were so striking that the scientists could match a dog with its owners out of a group of samples, solely based on the microbiota.

In a separate study, researchers found that the presence of a dog was associated with an increase in microbial diversity in the dust of the house that a dog calls home, and that many of the microbial species found in the household dust were also living in the dog owners' intestines. It seems that by bringing the outdoors in and by licking everyone and everything they can, dogs act as microbe delivery systems that equalize the microbiota across the household.

In both studies, cats didn't seem to influence their owners' microbiota very much, which is likely due to the behavioral differences between the two species. Dogs like to play and tumble with humans and they lick a lot. Cats? Yes, maybe, but usually only when they think that you're truly worthy of their attention. Cats also don't beg you to take them on a walk and, due to their tendency to run away for days at a time, they aren't brought outside as much as dogs. Cats and dogs are both great pets, but when it comes to the microbial gifts that pets bestow on their owners, dogs win fair and square. We'll take soil microbes over a dead mouse any day.

Bring on the Slobberfest

Like many parents and grandparents around the world, Nathan and his mom (from the anecdote above) had this notion that dogs, and dog slobber in particular, could make a baby sick. To some extent this is true. On rare occasions, dogs can pass on a disease to a child (or to anyone) because they can harbor all sorts of worms (heartworms, tapeworms, hookworms, etc.), other pathogenic bacteria, and viruses. However, these diseases are very rare among pets that are well looked after and that receive veterinary care periodically. Sure, if a dogs looks sick, has diarrhea or a skin rash or scab, it's probably a good idea to get the dog to the vet instead of letting your kid roll around with his furry friend, but there's a very low risk of catching an infectious disease from a dog that receives good care.

On the contrary, owning a dog that goes outside and allowing it to interact with children is actually beneficial for their health. Epidemiological research shows that kids that are exposed to dogs early in life have a decreased risk of developing asthma and allergies. A 2013 article published in the *Journal of Allergy and Clinical Immunology* summarized the results of twenty-one studies that aimed to figure out what factors contribute to the development of childhood allergies. What they found is that exposure to a dog during pregnancy or before the age of one decreases the risk of developing eczema (a skin disease) by 30 percent. In several other studies the presence of a dog (but, again, not a cat) is also associated with a reduced risk of asthma, decreasing the risk by about 20 percent. This recent information has surprised allergists around the world, who for years recommended removing pets from home to reduce allergies (although in certain Central and South American countries dogs have been used to cure asthma; see Chihuahuas Cure Asthma?, page 124).

Many people do develop allergies to pets, and the presence of a pet in the house can exacerbate the problem if a child is allergic to something else. In these cases it makes sense to consider finding another home for the pet. However, since studies show that the presence of a dog may prevent the development of asthma and allergies, unless Milo gets sick or someone develops an allergy to him, promoting contact between Rory and his four-legged friend is actually good parenting! Parents and grandparents everywhere please take note, though: buying a pet solely to decrease your child's risk of asthma is not a solid enough reason to own a pet. A dog is a big commitment, especially with a new baby in the house; they require a lot of attention, training, walks, and money. If you don't see yourself wanting this added responsibility, it might be a good idea to hold off on getting a pet for now, and let your baby play with a family member's or friend's dog instead.

The strong relationship between having a dog and the reduction of asthma and allergy risk certainly raises the question: What's so special about dogs? We've suggested that it's the microbes in the dirt that a dog brings into the house, but others remain skeptical, claiming that it could perhaps be something that the dog produces instead (this is a good example of the type of things scientists love to bicker about!). What settles the argument in favor of the dirt microbes theory is a study led by Dr. Susan Lynch from the University of California in San Francisco. This study collected dust samples from homes with and without dogs, and showed that upon exposing mice to the different dust samples, the mice that were given dust from homes with dogs were less likely to develop asthma. What's more, they looked at the type of bacteria in the dust samples and found a specific species, *Lactobacillus johnsonii*, associated with the improvement of asthma in mice. When they grew this bacterium in the lab and fed it to mice in the absence of any dust, they found that it lowered the risk of asthma, demonstrating that this and perhaps other species of beneficial bacteria, along with the dogs that bring them into households, are responsible for decreasing asthma risks.

These types of studies have important implications. If dogs transmit bacteria that make humans less prone to an immune disease, this implies that dogs carry around probiotic species that are beneficial for human health. What are they? Can they be grown in a lab and given to kids? We have a lot more to learn in this area, and scientists are certainly working on it. What *is* clear is that dogs and humans have a special partnership that goes beyond their loyal friendship. Dogs keep us dirtier, and as we have come to learn, kids benefit from this kind of exposure early on.

Dos and Don'ts

- ◆ **Do—** let your dog safely play and closely interact with your baby or small child. It's a good idea to take the dog to the vet right before the baby arrives, just to make sure your pooch is in good health. Letting your dog lick or be close to your baby is likely to decrease his risk of developing allergies and asthma, with the added benefit of providing companionship and protection, and teaching your child to be comfortable around animals.

- ◆ **Don't—** consider getting a dog (or any pet for that matter) simply to decrease a child's risk of asthma. Owning a pet is a lot of work and they deserve to be taken care of by committed owners that will provide food, veterinary care, and entertainment. Expose a child to a dog you know if you can't have one at home.

- ◆ **Don't—** shun cats just because they don't seem to offer us beneficial microbes. They make fabulous pets, too. However, cats are known to transmit parasites through their feces, so it's recommended to avoid changing the litter box during pregnancy and moving it to a place a baby can't access it.

CHIHUAHUAS CURE ASTHMA?

In many Central and South American countries there's a surprisingly widespread belief that if a baby or young toddler suffers from wheezing or allergies, buying a Chihuahua will ease the symptoms. This rumor has been around awhile (noted in medical journals in the 1950s) and has spread to the southwestern United States as well. As a result, many families of asthmatic children buy little Chihuahuas.

There are actually two versions of this story: one claims that the dog cures the disease altogether, while the other posits that the poor dog soaks up the disease from the human, becoming asthmatic in the process! What's even more unbelievable is that certain family doctors and pediatricians (especially old-school ones) have recommended this practice to patients, based on their own experience.

It would be an understatement to say that there's serious skepticism about this theory. It's true that Chihuahuas, as a short-haired breed, are considered hypoallergenic since they shed very little, but it's highly unlikely that living with a Chihuahua is going to cure someone's allergies. But hey, if you feel like snuggling in bed with an adorable little dog, go for it. It will certainly make you happier.

9: Lifestyle:
Microbe Deficit Disorder

Starved for Nature

Societal changes in the past century have dramatically shifted how we live. The type of work we do, the places and types of buildings we live in, what we do for entertainment, and our family dynamics—just to name a few—are very different than only three generations ago. Because of this, being a kid now is very different than thirty, sixty, and especially a hundred years ago. Without a doubt, many of these changes have been positive. For example, 96 percent of US children go to school now, compared to only 60 percent in 1913, and the infant mortality rate has gone from 150 deaths per 1,000 births to 5 deaths per 1,000, all in a span of one hundred years—both remarkable accomplishments in societal development. However, some of the changes brought about in the past century may not be as positive. Kids have always loved to play just like kids do today, but the activities they undertake and the places in which they play are vastly different. Inadvertently, these changes have resulted in a detachment between children and the outdoors, and this has had a major influence on microbial exposure in children.

For a real-life example of this, try a simple experiment: next time you're at a family reunion or any other event where several generations are present, ask the members of various generations what they used to do for fun when they were kids. Try to get them to think not only of the activities they used to participate in, but also where they played and the amount of time they spent outside. We presume the answers you get will be along these lines: "We got home from school, ate something, and were not back in the house until it was time for dinner," or "We would roam the neighborhood in packs of ten to twenty kids, climbing trees, building forts, chasing one another until it got dark," and our favorites, "Being in the house meant we were sick or grounded" and "I remember my knees were permanently dirty and scratched." Now compare that to, "I love to watch Netflix," "Videogames!" and "IM'ing with my best friend." It seems almost unnatural to think that the childhood memories that modern children are creating mainly happen indoors, away from nature, but this is the reality.

The statistics are truly disheartening: children spend half as much time outside as they did only twenty years ago; kids ages 8–18 spend a daily average of 7 hours and 38 minutes using entertainment media or screen time; and only 6 percent of 9–13-year-olds go outside by themselves. As unrealistic as it sounds, in England more children are now taken to emergency rooms for injuries incurred by falling out of bed than falling out of trees. It's no wonder that kids are so sedentary these days, when their idea of good entertainment comes from a screen and not from running outside and physically playing with other kids. The situation is so dire that health agencies in many countries have issued minimum physical activity requirements for children, something that only thirty years ago would have seemed like a ridiculous policy. What's worse is that out of fear for their safety, or fear of them getting dirty or injured, adults are constantly supervising children or even keeping them from going outside altogether!

As we have said in the previous chapters, our resident microbes are a result of what we physically interact with and the food we eat. It's truly worrying to think that millions of children are growing up mainly exposed to indoor microbes, like the ones growing on their Wii remotes or computer keyboards. For millions of years, children have grown up exposed to a substantial number of outdoor microbes and this connection has been broken in the past couple of generations—which coincides with the time it has taken for Western lifestyle diseases to skyrocket.

Such a Thing as Too Clean

One of the main reasons parents and caretakers today have an aversion to letting kids freely play outside is the notion that they can get sick from putting dirt or dirty objects in their mouths, or from being dirty for an extended period of time. This is an ingrained perception cultivated over decades—the idea that "dirty" inevitably means the potential for infectious disease. We spent generations avoiding harmful infectious agents in the environment and cleaning up our world. The World Health Organization defines hygiene as "the conditions and practices that help to maintain health and prevent the spread of diseases," and there are indeed many proven advantages to following hygienic practices—namely, the spectacular drop in childhood mortality. However, Western societies have taken hygienic practices to the extreme. The concept of cleanliness (often cited as next to Godliness!) is not necessarily associated with health benefits but with physical appearance, and our modern societies have never been so clean. Never have there been so many brands of soap, deodorant, toothpaste, razor blades, disinfectant, shampoo, lotion, and perfume. Being clean is our standard of living; clean feels good (don't shower, wash, or shave for a week, and you will no doubt agree).

It's important to keep in mind, however, that cleanliness does not have an advantage over hygiene in preventing disease. Cleanliness is a relatively new concept; it's been part of our culture for only a hundred years or so. Before the mid-nineteenth century in the US, regular baths and teeth brushing were not common practices, and neither was using soap. The first hygienic measures took place through an organization known as the Sanitary Commission, which originated during the American Civil War. It was very successful in reducing infectious diseases and deaths by promoting washing the sick, along with their bed linens and their rooms. Back then doctors and scholars were just beginning to accept the concept that germs transmitted diseases.

Our current cleanliness practices have become more of a cultural construction based on the idea that the cleaner you are, the better it is for you. As an example, people might be disgusted by a picture of an infected wound or someone covered in dirt. In reality, the wound is an actual threat to your health because it's infected, whereas dirt is not a threat at all, it only looks unclean.

Most people become even stricter about cleanliness when they're taking care of a baby. This makes sense, as it's a natural way to protect a baby and prevent infections, but our modern sense of how clean we should be is causing babies to be brought up too sterile. When Brett was a kid, every morning after school prayer (at a public school!) he had to stand in line for hand inspection. If there was dirt under your nails, you were wrapped hard on the knuckles with a ruler, then sent to the bathroom with a brush and you couldn't come out until they were clean. This story was pretty typical at that time, and although hand inspections don't happen like this anymore, the level of cleanliness in a child is still a reflection on how well he is taken care of, so keeping them squeaky-clean is considered good parenting by lots of people.

As a result, and with the advances in cleaning technologies, gel hand sanitizers hang from almost every diaper bag; toys and pacifiers are wiped clean with antibacterial wipes if they hit the ground; bottles and utensils are sterilized before every use; babies and toddlers are often not allowed to play in the dirt or sand, and when they are, they are wiped clean immediately. Phrases like "Ewwww!" "Yuck! Don't play in the mud!" or "Don't touch that bug, it's dirty!" have become second nature. Babies and children are prevented from following their innate nature to get dirty, and in doing so they're being shielded from the microbial exposure that's essential for their development.

Recognizing that we live too cleanly may not be hard, but learning how to differentiate a potential health threat from something that only looks filthy is not always easy, and in some cases it's not a black-and-white decision. As scientists, we've studied microbes that cause diseases for many years, but as parents, knowing all that we know, it *still* hasn't been easy to make decisions regarding microbial exposure. So we polled parents to find out what they most wanted to know—their most pressing questions and concerns—and then applied current scientific knowledge to provide answers (for an extended list, visit www.letthemeatdirt.com).

Cleanliness Q&A

1. When should children wash their hands? What kind of soap should we use?

Handwashing is, without a doubt, the best hygienic practice that we can follow to prevent contracting and spreading infectious diseases. It's been shown time and again that communities with good handwashing practices stay healthier, and no one should stop washing their hands just to promote more exposure to microbes. With that said, children do not need to wash their hands all day long. Handwashing should occur before eating; after using the toilet; after being in contact with someone sick, or, if the child is sick, before she touches other people; after touching garbage or food that is suspected to be decomposing; after touching animal waste or farm animals; or after being in places frequented by many people (public transportation, malls, etc.). Children do not need to wash their hands after playing outside, unless they are about to eat; immediately after they walk into the house; or after playing with other children, unless they are sick with an infection. Children should be outside often and should be allowed to be barefoot and to get dirty, and handwashing does not necessarily need to immediately follow these activities. The above list is certainly not exhaustive, but it is aimed to differentiate the types of exposure associated with the risk of infection from the ones that are not.

Regarding the soap question, the kind you use depends on personal preference, but it's best to avoid antibacterial soap. A Food and Drug Administration (FDA) committee found that antibacterial soaps provide no benefits over regular soap and water. Except for hospitals or places where additional medical hygiene is necessary, antibacterial soap doesn't have a place in everyday use, and the same goes for antibacterial sanitizers. Plain old soap and water is enough and an alternative sanitizer (like a gel sanitizer) should only be applied if there isn't a potable source of running water and soap.

This advice is certainly counterintuitive to common practices, and it may also be hard to follow since finding soaps without antibacterial agents in them can be harder than you think (most liquid hand soaps have them). A common antibacterial chemical used in soap, called triclosan (and its derivative triclocarban), is also found in deodorants, toothpastes, cleansers, and cosmetics. Triclosan has been used for about sixty years, but recent research questions its side effects and environmental toxicity.

Besides killing bacteria, triclosan has been shown to alter hormone regulation in animals and it might contribute to the development of antibiotic resistance in bacteria. Triclosan is classified as a toxic chemical for aquatic organisms and is known not to biodegrade easily. Big companies such as Johnson & Johnson pledged to remove triclosan from all their product lines by 2015 (at the time this book was written, they hadn't yet confirmed the removal). Triclosan has been banned in Europe since 2010. The Canadian Medical Association has suggested a ban on antibacterial consumer products such as triclosan, and the FDA states that the risks associated with the long-term use of antibacterial soaps may outweigh their benefits. Still, triclosan is currently considered a safe product for human use in many places and it's up to the public to refrain from using it.

2. Should I let people hold and touch my baby?

The answer to this question is ultimately a matter of personal choice, depending on how comfortable you feel about passing your baby over to someone else. With that said, research shows that social interactions, including physical contact, is one of the ways to maintain diverse microbial communities. In one study, biologists took samples over a long period of time from two groups of African baboons that lived near each other. These two groups had the same type of diet, yet they differed in one important behavior: one group engaged in social grooming and the other did not. Interestingly, the microbiotas of these two groups were different, and the baboons that groomed one another had more similar bacterial communities than the baboons in the group that did not. This study shows that it's not just the food we eat that determines the type of microbes that grow within us, but that social interactions like physical contact play an important role. Thus, limiting physical touch, a behavioral trait that characterizes humans as a species, likely limits the exchange of microbes between a baby and his surrounding humans.

If the fear of your child getting a disease is what prevents you from letting other people hold or touch your baby, there are ways to significantly decrease this risk. First, avoid having your baby around people that are sick with an infection; second, ask everyone to wash their hands before holding and touching a very young baby. Since letting a baby be touched is one of the ways he gets exposed to microbes, it makes sense to want to avoid the chance of infection, but physical contact with healthy people is not going to be dangerous for the child, and may even be beneficial.

3. If my child is sick with a cold, should I keep her at home to avoid spreading the illness to other children, or is it advantageous for kids to share infections to toughen their immune systems?

This is definitely one of those questions that doesn't have a black-and-white answer. Although there is robust evidence that exposure to microbes during early childhood can protect against certain immune diseases like asthma and allergies, there is no evidence that we need to be exposed to pathogenic bacteria or avoid hygienic practices to prevent immune diseases later in life. With that said, it's impossible to grow up without getting an infection. Suffering infectious diseases is part of being human, and especially part of being a kid. Preventing a child from getting sick at all costs will likely result in the type of behavior that also prevents a child from being exposed to many beneficial, non–pathogenic microbes. A child should not be bubble-wrapped out of fear of them catching the common cold or any other common pediatric infection.

While it's important not to constantly worry about a child catching a cold or other infections, it's also important not to let a child become a disease transmitter. If a child is sick, the best idea is to keep her at home, simply to limit the spread of disease. No one wants to see a child with a bad cold or worse, chicken pox, show up at a birthday party, although it won't terribly harm a child to play with a kid that has a runny nose in the playground. Plus, if a child is under the weather, why not let her rest at home, giving her the best chance at a speedy recovery?

The above answer, however, is given in the context of a Western society, where most people are vaccinated. Vaccines are an artificial way to expose children to harmful microbes without them getting terribly sick. It's only because of vaccines that children these days have a very low risk of catching serious life-threatening infectious diseases such as smallpox, polio, diphtheria, etc. Fifty years ago, allowing your child to play with a friend that had a fever could have exposed her not just to the common cold viruses, but also to meningitis, whooping cough, measles, and other serious diseases. If we were to live in a world where only a subset of the population was vaccinated, the advice here would certainly be different.

Likewise, if you have decided not to vaccinate your children, understand that your children are more likely to suffer serious life-threatening diseases, as well as to carry and spread them to others. Thus, it would be prudent to limit the contact your child has with other children when she gets sick. (See chapter 15 for more on vaccinations.)

4. What about kids touching dirty surfaces?

First, not all dirt is created equal, nor does it all pose the same risk of disease. It would be almost impossible to accurately know which dirty things have pathogenic bacteria and which don't by simply looking at them, but there are a few giveaways. If something smells bad, looks slimy, or looks inflamed (in the case of a wound), it's likely harboring nasty microbes. This is especially important with food, where disease-causing bacteria love to grow, so don't touch food that smells or looks like it's decomposing, and be vigilant about expiration dates and preparing and cooking foods properly. Kids shouldn't be allowed to touch wounds or bodily fluids in general, but especially if they come from someone who's sick.

For those of us who live in cities with lots of other people, it's a given that many of those people will have an infectious disease at any given time. Picture yourself with your children riding the New York City subway one afternoon. It's very likely that someone in the same train car as you has an infection, or that someone who rode the train immediately before you sneezed and left a lot of viruses to spread on the same window that your child eagerly touches while enjoying his ride. Does this mean that we should avoid trains or any other form of public transportation, or that we should frantically apply hand sanitizer every time we touch a heavily commuted surface? No, of course not. If we were so susceptible to disease transmission that a train ride in New York City would carry a dangerous risk, the human race would have been wiped out thousands of years ago. Our amazing immune system is strong and can deal with this type of exposure; however, it pays to follow hygienic practices in order to reduce the risk of infection in heavily populated areas. This means that it's a good idea to teach your children not to play on the floor in these places, nor to lick any surfaces, and to wash their hands (with regular soap and water) when they get home or before eating.

When children are out walking or playing in a green space, it's a different situation altogether, as the risk of getting infected with microbes that carry human diseases decreases drastically. Allow your children to touch anything they want (except animal waste), including dirt, mud, trees, plants, insects, etc. Don't act on the urge to clean them right after they get dirty, either; let them stay dirty for as long as the play session lasts or until it's time to eat. In fact, our children experience so little time outdoors compared to previous generations that it's ideal to encourage them to get dirty during the little time they have outside. Bring a bucket, some water, and a shovel the next time you're at a park or on a hike—it takes only minutes before most of them start making mud pies or decide to give themselves (and you!) a mud facial. If the dirt gets in their mouths, don't freak out; they'll soon realize that dirt doesn't taste all that good and likely won't develop a habit for it. Most kids have this innate desire to get dirty, but in their modern lives this needs to be nurtured. Do your child a favor and encourage him to play with dirt.

5. Is it necessary to sterilize milk bottles and, if so, until what age?

This may come as a shocker for many of us who grew up with the idea that babies should be given only sterile bottles, but the American Academy of Pediatrics no longer recommends sterilizing bottles used for babies of any age. It recommends only a stove top method (cleaning them in boiling water) or a dishwasher with a hot cycle when using the bottles for the first time or when the water used at home is not deemed safe to drink. If the water is safe enough to drink, it is also safe enough to use to clean bottles and nipples. The same goes for any utensil or plate used to feed babies solid foods, and for pacifiers and teethers: washing them with water and soap is enough. However, be aware that milk bottles and nipples need to be washed properly as milk residues get trapped in the nooks and crannies of bottles and their accessories, and bacteria can flourish there. A bottle brush is a good idea for a proper washing.

One recent study may provide parents with an incentive to become more lax about sterilizing bottles and food utensils. A group of scientists from the University of Gothenburg in Sweden analyzed data from over 1,000 children and found that children from homes that washed their dishes by hand were less likely to develop eczema (a skin disease) by school age. The study controlled for other factors known to decrease asthma and eczema risk, such as having a family pet or breastfeeding, strengthening the validity of their finding. This study suggests that a less-efficient dishwashing method promotes more exposure to microbes early in life, which has been shown to protect children from allergies and asthma. While we may not advise tossing out your dishwasher, perhaps doing dishes manually on occasion would be appropriate (and a good way to learn to appreciate your dishwasher).

6. How often should baby/child toys be cleaned and what should be used to clean them?

This question was a popular one among the parents we interviewed. It often came accompanied with suggested answers, such as "Every day or after every use?" or "With regular disinfectant or with bleach?" But really, it's not necessary to wash toys until they are visibly dirty or after a sick child has played with them.

As for what to use, soap and water is more than enough. Harsh chemicals such as the ones in disinfectants or bleach are not necessary for this type of cleaning or for cleaning the surfaces where children play, either. (This was one of the questions that made us realize how widespread the notion is that babies need to play and develop in a pristine environment.)

7. Are sandboxes unsanitary?

Kids love sandboxes, and it's not surprising to find a dozen children playing in one all at once, making them a popular playground spot that undoubtedly has a higher concentration of microbes than other playground features (there are a couple of studies showing this). This means that a child has a higher risk of contracting an infection in the sandbox than on the swings or slide. Does this mean that they should not go in the sandbox? Absolutely not! Sandboxes are great fun and the risk of contracting a disease from one is low.

However, parents and caretakers should follow hygienic practices, like handwashing, after using the sandbox. The other possible source of infection comes from the fact that a sandbox looks like a giant litter box to many animals (read: cats) and they will use it accordingly given the chance. For private backyards, covering the sandbox after use can easily prevent this, whereas at a public playground it would be wise to inspect the sandbox before letting a child use it. If animal waste is visible, scoop it out, along with a good amount of the sand surrounding it (most of us have changed a litter box at one point or another). If the sandbox looks like it's been used by all the cats in the neighborhood, don't let a child play there and contact the local authorities to have its sand replaced (cat feces can contain parasites which can then infect humans).

8. Should my child be allowed to put something in his mouth after it's been dropped on the ground?

In general terms, putting something that has fallen on the ground back in your mouth is just fine. However, not all ground surfaces are the same and common sense applies. If a child's toy falls on the subway or mall bathroom floor, it's a good idea to give it a rinse with soap and water first, but if it falls on the floor at someone's home or while out hiking, simply remove the visible dirt (and hair) and give it back to your kid.

In fact, a recent study by the same Swedish research group that reported the association between dishwashers and an increased risk of allergies, suggests that the best way to clean a pacifier that has been dropped is to put it in your own mouth first. In this study, 184 families were interviewed when their babies were six months old. Parents were asked the question: Does your child use a pacifier, and if so, does it get sterilized, rinsed in tap water, or cleaned by parents sucking on it? Surprisingly, they found that the sixty-five babies raised by parents that cleaned their pacifiers by mouth had a significantly lower risk of developing allergies at eighteen and thirty-six months of age. This small study remains to be replicated, but it seems that by sharing mouth microbes with their child, parents are strengthening their child's immune system and preventing the development of allergies. So instead of following the "five second rule" to pick something off the ground quickly, perhaps instead we need to follow a "five second rule" in mom or dad's mouth before returning a teether or pacifier to the child. (There may be a concern with parents passing cavity-inducing microbes to their children, but this appears to be an issue only with parents who are prone to tooth decay, which can be hereditary.)

9. Is antibiotic ointment necessary in treating scratches and cuts?

Not always. Cuts, scratches, and scrapes are part of being a kid and they happen all the time. If the wound is long or deep or the edges are far apart, or if it doesn't stop bleeding after a few minutes of applying pressure, seek medical attention. Otherwise, wounds simply need to be cleaned of dirt and debris by thoroughly washing them with soap and water (or immediately rinsing them with clean water, and washing more thoroughly with soap later, when available).

The recurrent use of a little dab of antibiotic ointment may not significantly alter a child's skin microbiota, but it adds to the unnecessary use of antibiotics, which leads to antibiotic resistance. To prevent developing an infection, keep the wound clean by gently washing it daily and avoid touching it by covering it with gauze or a bandage. If, after a day or two, the wound looks red and swollen, or is oozing yellow or green pus, consider using an ointment with antibiotics. If the redness around the wound expands or there are red streaks spreading from the wound, or if there's a fever, seek medical attention.

Luckily, most cuts and scratches heal rapidly on their own due to our immune system's ability to control infections.

10. Should I allow my child to eat unwashed fruits and veggies?

In most cases, you should wash produce. Fruits and vegetables are often consumed raw, which means that any contamination that occurred during farming or storage may come in contact with whoever eats them. The irrigation systems used to water many types of crops are known to contain dangerous pathogens, and washing fruits and vegetables is an effective way to significantly reduce the risk of foodborne diseases. Foodborne diseases, also known as food poisoning, are more likely to occur in certain groups of people, including children, the elderly, and pregnant women (they all have more vulnerable immune systems). The CDC estimates that about one in every six Americans get sick from food poisoning; each year 128,000 people are admitted to hospitals, and 3,000 people die from food poisoning. Thus, this is a serious risk that ought to be reduced by following hygienic practices.

Washing food that will be consumed raw is only one of the steps to prevent foodborne disease. Other important practices are to separate raw meat, seafood, and eggs from ready-to-eat food, to cook foods to the right temperature, and to chill or refrigerate perishable foods within 1–2 hours after purchasing them.

Another good reason to rinse fruits and vegetables is to wash out pesticide residues. Some people use a fruit wash solution for this purpose, although the European Crop Protection Association and the American National Pesticide Information Center both state that using these types of products is no more effective in removing pesticides than water alone.

The only type of situation in which we would consider it okay to allow a child to eat an unwashed piece of fruit is if it was grown in her backyard or garden, watered by rain or clean water from a hose, and free of pesticides. However, this shouldn't lead to the idea that it's okay to consume unwashed, store-bought organic produce because, contrary to popular belief, organic farming does not decrease the risk of food poisoning, although it does significantly reduce the levels of pesticides in crops.

Organic foods are often fertilized with manure, which can contain pathogens. For example, there was a major outbreak of *E. coli* O157:H7 (which causes serious diarrhea and kidney failure) in apple juice made with organic apples that had been exposed to cattle feces that contained this pathogen.

11. How can parents promote the development of a healthy microbiota through diet?

This is a good one to answer because there's no better way to influence the development of a diverse microbiota than through diet. Offering a healthy diet rich in vegetables and fiber is probably even more important than not being overly clean with babies and children. As we mentioned in chapter 6, when babies start eating solid foods they should be given a diet varied in vegetables, fiber, and fermented foods. A child can be exposed to many good sources of microbes while playing and interacting with people, but if these microbes are not fed the right foods, they won't flourish in a child's gut. If a child's diet is mainly based on refined carbohydrates (white flours and sugar) and high fats, his digestive system will digest and absorb most or all of the nutrients in the upper part of the digestive tract, leaving little nourishment for the vast numbers of microbes inhabiting the large intestine farther down. The microbiota in the large intestine feed on fibers and foods that are somewhat resistant to digestion in the upper part of the digestive tract and if none of that makes it down they will starve and diversity will decrease.

While offering babies lots of vegetables, legumes, fiber, and fermented foods is a good idea, convincing a two-, three-, or four-year-old to eat their carrots and celery is a whole different game. Once babies realize that they can make their own decisions, they will try to get only what they want, and upon tasting french fries or ice cream they will undoubtedly shun anything else, especially if its green. This is when teaching them good eating habits becomes extremely important and extremely hard at the same time. Even the most dedicated parent can easily give in after the thirtieth time their toddler asks for candy, especially if this happens late in the day when the patience levels are low.

Claire found that her daughter became a lot less resistant to eating all the healthy stuff when, one day, when Marisol was three and a half, Claire told her that she had a huge collection of little bugs in her tummy. She made the story very elaborate and whimsical; the bugs all had different colors and shapes, they sang songs, had parties, and simply loved living in Marisol's tummy. They actually called her tummy their home, and they were the happiest little creatures that ever lived. They also had superimportant jobs to do, like chopping up all the food she ate into really small pieces so it could reach the rest of her body to make her grow. Her little bugs were also her poop factory and they made sure that she recovered when she got sick, too (all only slight exaggerations from the real facts!). Claire also said that these bugs were always hungry because of how busy they were and that Marisol's job was to feed them every day. Without food they would starve, get supersad, and even die. But her bugs did not like ice cream, candy, hamburgers, or french fries; they loved lentils, broccoli, kefir, beans, carrots, and tomatoes instead. This, Claire told her daughter, is the reason why we need to eat vegetables and other foods that aren't as tasty as cupcakes. "It's not for you" Claire said, "it's for your bugs. You're their home, and it's your job to feed them." The change in her daughter's attitude towards the foods she didn't like was almost immediate. She gave her tummy bugs pet names, drew elaborate pictures of what she imagined they looked like, and agreed to feed them well.

It's been almost two years since then and Claire's daughter continues to eat her vegetables. To her, the story became engrained, something matter-of-fact, just like the fact that there are four seasons in every year, that Sunday comes after Saturday, or that Santa lives in the North Pole.

As Claire's children grow, she will likely have to modify the message to one with more realistic tones and details, but the core message is the same. The microbiota is a forest that we carry inside us and it is our lifestyle choices that determine whether this forest is stable and balanced or fragile and hungry. So it's important that children eat a diet that promotes a diverse microbiota and to establish good eating habits that will hopefully last for many years to come. Teaching them to eat well, as well as teaching them that being squeaky clean is not the right way to go about life are two crucial messages that children should get. It's those early years that matter the most when it comes to microbial exposure and the development of their immune system, so it's well worth the effort to change our preconceived notions regarding diet and cleanliness to promote a healthier future for them.

PART THREE

Collateral Damage

10: Obesity:
The World Is Getting Heavier

Body Weight and the Microbiome

Jack Sprat could eat no fat,
his wife could eat no lean.
And so between them both, you see,
they licked the platter clean.

We all know that increased body weight, especially in children, is a huge problem (pun intended). The statistics are downright scary: childhood obesity has more than doubled, and has even quadrupled in adolescents in the past thirty years. Between one-quarter and one-third of all American children are either overweight or obese—and the numbers continue to increase. There is also, not surprisingly, a direct correlation between childhood obesity and adult obesity. The average American woman now weighs the same (166 pounds) as the average male weighed in the 1960s (yes, men are equally guilty of major average weight increases). The problem is that packing on the pounds translates into serious health problems, including cardiovascular (heart) disease, strokes, diabetes, and cancer. Most health experts agree that it's the biggest epidemic facing the health of the developed world, and it has happened rather suddenly (within the past three decades). This time span suggests that it isn't due to a recent mutation in the human genome, but that something in the environment has changed to cause this sudden surge in human body weight. Ask any expert why this is happening and they will cite three factors: a change and increase in our diet, a decrease in our activities, and some unlucky genes.

Decreased activities are definitely a lifestyle change. Ask your parents or, even better, your grandparents what their daily routine was when they were your age. Answers might include hard physical labor on the farm, working outside, walking often, etc. Better yet, ask them what they did as a kid. Their daily activities probably would have consisted of a lot of running around, playing outside, bike riding to and from school and friends' houses, soccer, baseball, climbing trees, jumping rope, and so on. Now look around at kids these days. They are driven to school and they spend their spare time (and even class time) on computer screens, which all adds up, comparatively, to very little exercise.

Besides inactivity, the other factor affecting body weight is what we eat, and how much we eat. There has been a massive shift in our diet from unprocessed foods rich in vegetables and fiber to heavily processed foods with high levels of sugar. Corn syrup is a very inexpensive sweetener, and has found its way into many foods that were never before as sweet as they are now. High-calorie foods are cheap (especially fast food), and we consume a lot more of them than we used to. The average size of a bagel, cheeseburger, soda, or blueberry muffin has more than doubled in only twenty years, and children are getting used to these portions and adjusting their

appetite accordingly. In an effort to curb the mindless intake of liquid calories, whole cities have passed laws to limit the size of sugary
beverages that businesses can sell—a first for humankind.

So what actually governs how we process our foods and convert
them into energy? We have accepted that diet and exercise are the
main factors that influence weight, but is it as simple as that? This
formula doesn't explain how many people don't lose weight even
when they follow a strict diet and exercise regime. You've probably
guessed by now that the microbiome plays a major role in all this.
We know that diet changes result in microbiota changes. However,
could changes in gut microbes be responsible (at least in part) for
the obesity crisis?

We're now learning that even early life microbiota can have a
profound effect on a child's weight. In chapter 3, we saw that if a
pregnant woman gained more weight than normal, the child was
much more likely to become overweight. If a mother smokes, her
child is also more likely to become obese (this hasn't been directly
linked to the microbiota . . . yet!) Children born by C- section are at
a higher risk for obesity than vaginally delivered children. Children
fed infant formula are twice as likely to become obese than those
who are breastfed (see chapter 5). All of these events directly affect
the microbiota, and the gut microbiota play a major role in regulating energy extraction and metabolism from ingested food, affecting
both weight gain and loss.

Fat Mice

One of the most satisfying things in science is when fairly simple
experiments have obvious outcomes. A few such simple experiments

conducted by Dr. Jeff Gordon's group at Washington University in St. Louis and by other labs were fundamental in demonstrating once and for all that the microbiota have a major effect on body weight. Germ-free (GF) mice have 40 percent less body fat than normal (microbe-containing) mice, even though the GF mice consume 29 percent more calories than the control mice; that is, they eat more yet weigh less. GF animals can even be put on a high-fat diet and they're protected from obesity. However, if the GF mice are colonized with fecal microbiota from normal mice (mice are copro-phagic, which means they like to eat their poop, making the experiment pretty easy), the newly colonized mice increased their body fat by 60 percent in two weeks, just by gaining gut microbes. This alone shows that microbiota really affects body fat.

To take it a step further, they colonized GF mice with feces from obese mice, and the newly colonized mice gained much more weight than GF mice colonized with microbes from a normal weight mouse. From this experiment we can see that the microbiome from obese animals is more efficient at harnessing energy from food than the microbiome from animals with a normal weight. That a simple fecal transfer can seriously affect body weight is a groundbreaking finding.

Of Mice and Men

Although we can't ethically repeat the above experiment in germ-free people, several studies indicate that similar things occur in humans. For example, using human twins in which one was obese and the other wasn't, Dr. Gordon's group found that by transferring the human feces into GF mice, the mice that were fed microbiota from the obese twin grew heavier and gained more body fat than those that received microbiota from the lean twin. They also found that the microbiota from the obese donors was less diverse in its microbial composition compared to that of the lean donors, and this was also true in the colonized animals. On an encouraging note, they discovered that if they transferred in a set of microbes from the lean donors, they decreased the weight gain in the animals with the obese microbiota; in other words, the lean microbiota won over the obesity-inducing microbes. Studies of overweight children also found differences in their microbiota compared to that of normal weight children, and showed that that these changes preceded actual weight gain. All these studies suggest that the bugs in our gut really do affect our weight. But how?

This is where things get more complex, and unfortunately we have to leave the simplicity and beauty of fecal transfer experiments behind. Several reasons have been proposed as to how and why microbes affect our weight, but at this time they're really just theories and, frankly, we don't yet know exactly how this all works. We know the microbes do the bulk of the hard work in breaking down our food (humans don't even bother having the genes that produce the enzymes needed for digestion of certain foods, as we know the microbes have it covered). As shown in the GF mice studies, the obese microbiome seems much more efficient at harnessing the energy from food. These microbes have more enzymes dedicated to food breakdown and energy harvesting than microbes from lean people. Some of these breakdown products trigger a hormone release in our body that affects whether we feel full or not (from a microbe's point of view, what a great way to get more food!).

Another potential reason involves inflammation, a condition that occurs in obesity, and which is thought to induce obesity-associated diseases like type 2 diabetes and insulin resistance. Inflammation is an immune response to pathogens or tissue injury. In the context of obesity, inflammation is thought to occur because high-fat diets trigger an increase in gut permeability, making the gut leakier and allowing microbes or microbial molecules to pass through the gut wall, which in turn triggers a general inflammatory state (this is discussed in more detail in chapter 12, where we talk a lot about gut permeability and gut diseases).

Many necessary details are still lacking in order to explain exactly how certain microbiota compositions favor weight gain and obesity, but the general concept that's emerging is that the gut microbiota modulate the body's ability to absorb energy through various methods, all as a result of the by-products of food degradation by the microbes.

A Microbiota Diet

So, knowing all this, can we tweak our microbiota to affect our weight? Although tempting, nobody has done the experiment yet in which you ask your thin spouse for his or her feces for a transplant (think Jack Sprat). However, attempts are now being made to directly alter the gut microbes, as well as the diet, that instigate changes in the microbiota. Prebiotics (dietary fibers that our body can't digest but that feed and stimulate particular microbes) have been given to healthy humans, and it turns out that they reduce hunger and make one feel full. This is probably because these prebiotics are modified by the microbiota, and they then affect the body's production of hormones, preventing a hunger signal shortly after consuming them.

The typical processed meal associated with obese individuals, on the other hand, contains only nutrients that are digested by human enzymes in the small intestine. By the time this meal reaches the colon, where most microbes live, there are very few food sources for them, which is thought to trigger hunger signals despite the body having obtained sufficient calories. Thus, it seems that in order to feel full you must not only feed yourself, but also your microbes.

Studies regarding infants and prebiotics are just getting under way, and it appears that they do cause an increase in beneficial microbes in young children. Some infant formulas now contain prebiotics in addition to probiotics.

Likewise, there have only been four randomized control studies done on probiotics in humans regarding body weight changes, and the data were inconclusive due to the small sample size. However, as we learn more about how microbes break down food, and the various roles they play in lean and obese individuals, it's quite likely that in the future we'll have probiotic-like mixtures of microbes that can be taken to decrease weight gain.

For now, we do know that when obese humans are put on a fat-restricted or carbohydrate-restricted low-calorie diet, a beneficial shift in their microbial population occurs, moving it towards a microbiota composition that promotes less weight gain. These changes occur rapidly, usually starting within twenty-four hours. It's also been shown that the physical responses of overweight adolescents to diet and exercise weight loss programs depends on their microbiota composition prior to treatment. This probably explains why some individuals are more successful on diets than others. It also again emphasizes how important the microbiome is in affecting body weight.

Antibiotics and Childhood Weight

As we saw in chapter 7, antibiotics are very good at killing microbes, both good and bad. And while these drugs remain wonderful at controlling serious bacterial infections, we must discuss a major dark side to them: they seem to promote weight gain, and their use may be directly contributing to the obesity epidemic.

Approximately seventy years ago, veterinarians made the observation that using antibiotics in subclinical doses (amounts that are less than would be used to treat an infection, but that will still affect some microbes) caused animals to gain weight by 10–15 percent. This effect is seen in pigs, sheep, cows, poultry, and even fish. It's made a huge difference in massive farming operations, which dose their livestock with antibiotics in order to get more meat off their chattel. This practice has become a cornerstone of agriculture in North America, and it now accounts for more than 80 percent of antibiotic consumption. However, this practice has dramatically increased the rate of antibiotic resistance, which is a major issue. It also leads to large amounts of antibiotics entering the environment. Europe has wisely banned the use of antibiotics as growth supplements in animals, but the United States and Canada stubbornly refuse to follow suit.

Initially it was thought that the antibiotics might control infectious microbes, thereby decreasing infections and allowing the animals to grow more quickly. However, the reason now appears to be more complex, and it's related to microbiota changes. Studies of animals showed that subclinical doses of antibiotics, no matter what kind, do indeed alter the microbiota to a population more conducive to weight gain, including an increase in energy-harvesting microbial genes. Several experiments also suggest that the weight gain is more pronounced if the antibiotics are given early in life, rather than later, hinting yet again at the critical role of early-life microbiota.

At the risk of offending proud parents, what do the results regarding pigs, sheep, cows, and chickens have in common with antibiotics and children? Recent evidence suggests quite a bit (plus common sense dictates that if it happens in so many diverse animals, it would logically affect humans, too—we are just animals, biologically speaking). Some compelling data show that the states in the US with the highest antibiotic usage also have the highest obesity rates. In a large Danish study involving more than 28,000 mother-child pairs, antibiotic exposure during the first six months of life was associated with an increased risk in the child being overweight at age seven, especially if the mothers were not overweight. In a Canadian study, antibiotics administered in the first year of life increased the likelihood of a child being overweight at nine and twelve years of age. The list of studies goes on and on, and they all overwhelmingly point to the fact that antibiotics, especially given early on, affect the microbiota, which in turn increases weight gain and risk of obesity.

We don't know as much about the effects of low-dose antibiotics in humans. The animal data are extremely convincing though, and presumably the high levels of antibiotic use in both society and agriculture suggest that even children who don't directly receive antibiotics may be inadvertently exposed to smaller doses that could still affect their weight. These exposures could come from environmental sources such as water—remember, tons and tons of antibiotics are used every day in agriculture, and consequently end up in our groundwater—or even from eating meat from animals raised with antibiotics.

Malnutrition

In direct opposition to the obesity epidemic that burdens wealthier nations, malnutrition continues to be a major problem in poorer areas of developed countries and worldwide (although obesity is surging in some of these areas as well). Malnutrition has major detrimental effects on a child's physical and mental development, including stunted growth and even impairment in brain development. Historically it was assumed that malnutrition was the result of a lack of calories, and the solution was to simply provide more food. However, this solution often does not work (it has been tried many times by feeding children in impoverished areas without success). A study done a few years ago showed that if the children were treated with antibiotics first, then many more of them gained weight, hinting at the role of the microbiota. Experiments have been conducted in which feces were taken from Malawian twins, one of whom was extremely malnourished and the other not, and transferred into germ-free mice. Similar to the results of the obesity studies discussed earlier, it was found that this fecal transfer also transferred the malnourished characteristics to the mice, which strongly supports the idea that the microbiota has a large role in malnourishment.

Work done in our laboratory has also confirmed the role of microbiota in moderate malnourishment. In an effort to develop a realistic animal model with which to study this major worldwide childhood problem, we fed mice two different diets that contained an equal numbers of calories, but were either rich in protein and fat (typical of a Western diet) or high in carbohydrates (typical of a developing country diet). As expected, changing the diet alone was not sufficient to mimic the features of malnourishment. However, we know that children in developing countries frequently live in a less sanitary environment, so are often exposed to feces. We also know that these children have more microbes in their small intestine (just below the stomach) that resemble the microbiota normally found lower down in the large intestine. These children presumably acquire these microbes orally via fecal contamination in their water or other sources.

Based on this, we fed mice feces from other mice that were on the two diets, and found that, remarkably, all the features of a malnourished child were found in the mice with a high carbohydrate diet, as long as they also were fed feces. We were also able to identify select microbes in the feces that caused this effect (we tried many, including probiotics, but only specific ones had the effect we were looking for). This suggests that certain microbes play a major role in malnourishment and that we finally have a good animal model with which to study malnourishment. Hopefully we can use this knowledge to develop therapies for this major global problem in the future.

Anorexia Nervosa

To be the thinnest. That is the tragic and tormenting goal of an increasing number of young girls and boys, mainly in affluent cities around the world (although it's starting to become more widespread). Anorexia nervosa has been called a silent epidemic because there's little awareness about this disorder, despite the fact that it has seen an increase of more than 50 percent in the past five years in North America and the UK. Anorexia nervosa (also called just anorexia) is a neurological condition that is characterized by self-starvation. It often occurs with other neurological issues such as depression (up to 80 percent of people with this disease also suffer major depression) and anxiety (75 percent have anxiety disorders). It's most common in adolescents, affecting 3 million Americans.

Unfortunately, anorexia has serious side effects on the heart as well as the entire body, and has a tragic 5 percent fatality rate, the highest death rate of any psychological disorder. Treatments for anorexia always include dietary interventions, but they're not always effective and relapses often occur. Recently, it has been suggested that the microbiota may be involved, based on two lines of reasoning: 1) the neurological involvement in the disease (which microbes affect; see chapter 14), and 2) the major weight-loss issues. A few small studies have indicated that the microbiota in patients with anorexia is different than that of control subjects.

A recent study out of North Carolina looked at fecal samples from sixteen anorexic women when they were admitted for treatment, and again when they were discharged (having reached 85 percent of normal body weight). The study found that the microbes were quite different between sample times, and that the microbe population in the women when they were admitted was not nearly as diverse as when they were released (although it still didn't reach the diversity seen in healthy people). As the patients were treated and gained weight, the study also found that their moods improved, and it discovered a correlation between microbiota diversity and the presence of certain microbes and a decrease in depression and anxiety. Again, this study doesn't prove that microbes cause anorexia and its associated depression and anxiety, but a strong correlation can be made. These types of studies will pave the way to more extensive analysis and ultimately determine whether microbes can affect the outcome of this tragic disease.

Given the role microbes play in food metabolism and weight gain/loss, and their impact on depression and anxiety (chapter 14), it's most likely that microbes will play a central role in managing this disease in the future.

Dos and Don'ts

◆ **Do—** avoid unnecessary exposure to antibiotics during pregnancy and early childhood. Antibiotic usage is increasingly associated with obesity, and, just like farm animals, our children are gaining weight at a much faster rate, setting them on a course for obesity and all its problems. On the other hand, antibiotics are a wonder drug for serious bacterial infections, and need to be used in certain cases. If your child is treated with antibiotics, you should consider various measures to promote the health of the gut microbes after the treatment is stopped. This could include breastfeeding, prebiotics, probiotics, and a varied diet rich in plant fibers.

◆ **Don't—** let your children spend their days in front of electronics. Get them out of the house, and promote physical exercise through walking, a trip to the playground, or an organized sport or other activity. And don't let the seasons be an excuse to keep your family indoors: swimming is a great way to keep cool during the summer and ice-skating can warm up your toes during winter months.

◆ **Do—** purge your kitchen cupboards of junk food and stock your shelves with healthy foods. Throw out those sweetened beverages, too, opting instead for plain water. By eating healthier you're not only treating obesity, you're preventing it by giving the microbes in your lower intestines something to eat so they don't send you signals to keep eating!

◆ **Do—** start reading the labels on the meat, eggs, and dairy you buy, and opt for products from animals not treated with antibiotics. Sometimes these products cost a little more, but it's worth it in the long run, especially since so much is still unknown about the effects of the antibiotics we consume through our food.

◆ **Do—** engage your children in the story of their gut. Even at a young age, your child can understand that there are good bugs in his tummy and they work hard to keep him healthy. Making our children responsible for their bodies and health at a young age is a great step towards a long life of good habits. Plus, they might be inclined to eat their vegetables because they want to, not just because you told them to.

THE 5210 DIET

Unfortunately, despite what you read on the Web, there's no magic bullet for dieting. We can't control our genetics (blame your parents), but we can control diet and exercise, and these are the two factors we have to focus on, especially as parents. Increased physical activity and decreased screen time are critical for both burning calories and helping the body develop (we know that exercise promotes a favorable microbiota). Eating healthy foods with plenty of plant fibers and avoiding sugar, such as that found in sweetened beverages, is also key to a healthy diet.

A dietary and lifestyle program called 5210 promotes just that. It's easy to remember the guidelines: 5210 suggests that every day your child should eat at least five fruits or vegetables, spend two hours or less on screen time, have at least one hour of physical activity, and consume zero sweetened beverages. It's also nice because you can hold up the number of fingers on one hand for each category to help your child count. The program is equally applicable to encouraging a healthy lifestyle for adults (although many of us would have to quit our jobs to avoid all that screen time).

The fact is, if you can maintain your child's weight in an optimal range for their age and height, you're doing them a huge favor for later in life, both in terms of maintaining a healthy weight, but also in terms of preventing major diseases such as diabetes, cancer, cardiovascular disease, and stroke.

11: Diabetes:
Microbes Have a Sweet Tooth

A Disease on the Rise

Glucose is the most widely used sugar in living organisms and is our body's main energy source. It's taken from our blood by our cells in order to energize them. However, glucose cannot get into the cells by itself. The pancreas releases a hormone called insulin right after we eat, and it attaches to the cells just like a key attaches to a door. Insulin signals the cells to start absorbing glucose and in this way it regulates sugar levels, keeping them from becoming too high or too low. An excess of glucose in the blood is not good, and if these levels remain high for a prolonged time, it can turn into a disease called diabetes mellitus (*mellitus* means "honey-sweet" in Latin), but we usually just refer to it as diabetes.

There are three main types of diabetes: gestational diabetes, which occurs during pregnancy with no prior history of the disease; type 1 diabetes, in which the body destroys specific cells in the pancreas that produce insulin, so cells are not stimulated to take up glucose; and type 2 diabetes, which causes the cells to become resistant to the effects of insulin—this particular form is closely linked to obesity. All three types of diabetes result from high blood glucose levels because cells are not able to absorb the blood glucose for various reasons. The hallmark symptoms of diabetes are frequent urination (because the high sugar levels pulls more water out of the body and into the urine) and a marked increase in thirst (since the body is trying to replace the fluids that are being lost). High blood glucose levels can cause serious long-term complications that include cardiovascular disease, kidney failure, foot ulcers and amputation, strokes, and blindness.

It's estimated that more than 380 million people worldwide have diabetes (there are an estimated 100 million cases in China alone), and this is expected to grow to 600 million by 2035, mainly due to type 2 diabetes, associated with increased obesity rates. The disease is thought to have killed over 5 million people in 2013 (to put this in perspective, HIV kills about 1 million people yearly, and a total of about 54 million people die worldwide each year), and up to one-third of the population in some areas of the world suffer from diabetes. Given that diabetes involves sugar uptake, which comes from food being digested in the gut, and also involves the host immune system, it's no surprise that the microbiota is increasingly thought to play a role in this disease.

A Sugarcoated Pregnancy

It's estimated that between 2–10 percent of pregnant women temporarily develop gestational diabetes. It resolves almost immediately after birth, but about 10 percent of these women will go on to have type 2 diabetes later in life. Once it's detected, gestational diabetes can be managed with diet, and sometimes insulin. The trick to detecting it lies in identifying an elevated glucose level, hence the urine and blood tests for sugar that are administered during pregnancy. For the glucose tolerance test, a mother-to-be must drink a horrible-tasting bottle of glucose solution, and precisely two hours later, her blood is sampled to see how the body handled this sugar load. When pregnant, there's a significant increase in the body's energy production (and consumption) in order to feed the developing fetus. Although it's not known exactly why some women are unable to control their glucose levels during pregnancy, it's speculated that the placenta somehow affects the body's sensitivity to insulin.

There's limited information about how gut microbes might affect gestational diabetes. Nutritional counseling and proper diets play a role in controlling this disease, which would of course also affect the microbiota. In one study of 256 women in their first trimester of pregnancy, women who received both nutritional counseling and standard probiotics (*Lactobacillus rhamnosus* and *Bifidobacterium lactis*) had the lowest rates of gestational diabetes, with decreased blood glucose levels both during pregnancy and for a year after.

We're still in the early days of defining the role of the microbiota in this disease, but the microbial contribution to energy production is well established, so in the future there may be ways to optimize this relationship in order to decrease the risk of diabetes in pregnant women.

Finger Pricks and Insulin Pumps

The hormone insulin is produced by specialized cells in the pancreas (called beta islet cells) and is then secreted into the blood, where it promotes glucose uptake by cells in the body. In some people, the body's immune system attacks the beta cells (a type of autoimmune reaction), destroying them and thereby stopping normal insulin production. When this happens, you have type 1 diabetes (T1D).

T1D is usually diagnosed in people younger than thirty, so historically it's been referred to as "juvenile-onset diabetes." It's one of the most common metabolic disorders in children and young adults. The prevalence of this disease has doubled in the past twenty years, and is set to double again by 2020. In Europe, it's increasing 3–4 percent per year, with the fastest rate of increase in children less than five years old.

Fortunately, patients with T1D can have a nearly normal lifestyle by regularly checking and monitoring their blood glucose levels with finger pricks and injecting insulin every day, or by having an insulin pump surgically implanted. A promising new therapy involves transplanting in new islet cells, which can then produce the needed insulin.

Like many diseases, there's a genetic component to T1D, and several genetic markers have been linked to increased susceptibility. However, less than 10 percent of those who have these susceptible markers will develop the disease, which hints at—you guessed it—environmental factors such as microbes playing a role. We also know that the rapidly rising rates cannot be explained by genetic changes alone, as humans just can't genetically change that quickly.

Other clues that suggest microbiota involvement include: the increased risk of having the disease if one is delivered by C-section; dietary changes early in life; and perhaps antibiotic use (animal data convincingly show that antibiotics increase risk, but it hasn't been proven in humans yet). The theory that breastfeeding decreases the risk of T1D remains controversial.

The microbiota of children with T1D is different than the microbiota of those who don't have it—it's less diverse and less stable, and lacks the microbes that produce butyrate (an anti-inflammatory molecule that improves gut health). These differences are also seen in prediabetic children, indicating that microbial changes precede the disease. Although the number of studies is still small, using animal models of this disease shows very strong evidence that gut microbes play a role, as we've seen the microbiota undergo changes in diabetic animals. Interestingly, antibiotic treatment of mice that are genetically prone to this disease actually protects against T1D.

As we will see in the next chapter, gut permeability (or gut leakiness) is an issue of many intestinal diseases. However, it also seems to contribute to T1D. People (and animals) who have the genes that make them more susceptible to T1D also have increased gut permeability. How this might contribute to diabetes is not yet known, but a leaky gut may allow microbial molecules from the gut to get through and somehow affect the body's immune response to insulin.

Early-life diet also seems to play a large role in T1D by modulating the body's immune system and encouraging its attack on its own beta cells. If infants who have the genetic risk factors for T1D are weaned on extensively broken down or hydrolyzed casein formulas, they have a decreased risk of developing T1D by age ten. On the other hand, if infants have a short breastfeeding period and are then fed cow's milk in nonhydrolyzed formulas, they have an increased risk of T1D later in life. Furthermore, in mice that are at risk for T1D, a gluten-free diet had a dramatic reduction in the disease, as well as an accompanying microbiota shift. How all this ties in to affecting the body's attack on its islet cells still isn't exactly understood. But we know that the body's immune system is responsible for this autoimmune attack, and that early-age microbiota play a key role in the immune's system development, which may contribute to T1D. However, microbiota also affect gut permeability, which could also affect disease. Unfortunately, we also don't yet have reliable data on whether probiotics could influence T1D later in life.

The Western Diet: A Life Too Sweet

As we saw in the previous chapter, due to high-calorie diets and lack of exercise, people these days are gaining too much weight. One of the most direct consequences of obesity is type 2 diabetes (T2D). It's actually quite difficult to uncouple the two diseases as they generally go hand in hand. T2D begins with the body's cells becoming resistant to insulin, which makes it not as effective at triggering glucose uptake. As time passes, the body may also decrease insulin production, and the liver may increase glucose production. Together, these cause increased blood glucose, leading to T2D and all its awful side effects, such as foot amputation and blindness.

T2D now accounts for more than 90 percent of diabetes cases in adults. Originally thought to be a disease that only affected adults, with the surge in childhood obesity, T2D is now appearing in children as young as three years old. We've also discovered that children born from mothers with gestational diabetes have an increased risk of T2D later in life.

Not surprisingly, just as the microbiota is linked to obesity, gut microbes play a central role in T2D. If feces taken from obese mice are transplanted into GF mice, they develop higher insulin resistance. Several microbiome studies have been done on humans with T2D, and the results show that the changes to their gut bacteria are very similar to those seen in obesity, which is certainly not surprising. Again, as we saw with T1D microbiota, there's a decline in butyrate producers; presumably there's a lack of butyrate as well, and a corresponding increase in microbes that could cause diseases.

A major role of butyrate is to dampen inflammation, and continual low-grade inflammation is a hallmark of both obesity and T2D, so lacking the organisms that make butyrate may contribute to these diseases. Feeding butyrate directly to mice was shown to improve their insulin sensitivity. Butyrate also decreases gut permeability, which may help prevent inflammation by keeping pieces of bacteria from passing across the gut wall and triggering inflammation.

There's increasingly good data to suggest that the microbiota associated with glucose tolerance play a role in the early stages of T2D. Some scientists have gone so far as to say that microbiota analysis could be used as an early diagnostic for T2D, since some of the changes occur before the full-blown disease. However, there are significant differences in gut microbe composition in different areas of the world due to diet, genetic background, age, cultural differences, etc. For example, when the gut microbes of European women were compared to those of Chinese women, the major differences made finding common marker microbes difficult. However, in both populations changes were noted in the microbiota of those with decreased insulin sensitivity, and in both groups there was a decrease in butyrate producers and an increase in potentially pathogenic microbes.

Can we manipulate the microbiota to affect the rates of T2D? There are several studies that suggest this could be possible. In one series of experiments, feces from lean male donors were transferred into obese males with poor insulin sensitivity, and six weeks later they found a significant improvement in insulin sensitivity in the recipients. They also saw an increase in microbial diversity, as well as an increase in butyrate producers. However, the effect seemed to depend a lot on the particular fecal donor, as not all lean donors had the same effect. These studies suggest that once we figure out the right bugs to transplant, this might be a viable therapy to decrease T2D.

Diet is another obvious, promising way to change the microbiota. In one study, six obese volunteers with T2D were put on a strict vegetarian diet, which improved their insulin sensitivity, and their microbiome shifted to a more regular composition. Likewise, consuming probiotic yogurt for six weeks led to a marked improvement in T2D patients, causing their circulating glucose levels to decrease. Probiotics could possibly be used at treatment, since they're known to tighten up gut permeability and decrease inflammation, which could help resolve the disease.

Results using diet, pre- and probiotics, and metformin treatment of T2D (see Drugging the Bugs—Metformin, page 170) are certainly causing pharmaceutical companies to sit up and take notice. In the future, we'll likely see more microbiota-altering therapies developed, not only for T2D, but for several diseases of the Western world.

Dos and Don'ts

- ◆ **Don't—** let your diet—and that of your children—get out of control. The complications of obesity and T2D are terrible, and they last a lifetime. Eating healthily and including foods that encourage diverse and healthy microbiota are important, not only for diabetes, but for many health-related problems throughout life.

- ◆ **Do—** have your blood sugars checked regularly during pregnancy if you're deemed high-risk, which includes excessive maternal weight gain, a large baby, excessive amniotic fluid, or being older than thirty-five. If undetected, gestational diabetes can lead to a very large baby, which can then lead to a C-section and/or obstetrical trauma (damage) to the mother. After delivery, these babies' sugar levels plummet, requiring serious medical attention, such as intravenous sugar intake. Gestational diabetes can be controlled if detected, but it has to be detected first.

DRUGGING THE BUGS – METFORMIN

Metformin is a drug commonly used to treat T2D by lowering blood glucose levels. Despite being approved for human use and being used extensively, exactly how this drugs works is not known, but there are strong hints that it may act via gut microbes. If the drug is delivered directly into the blood (intravenously), therefore bypassing the gut, it doesn't work. Also, the drug isn't effective in mice treated with antibiotics. Metformin causes a profound shift in the gut microbe composition to a healthier profile, and at least one of these healthier microbes can be given directly to mice in order to decrease T2D.

A recent large international study showed that T2D patients taking metformin had a different microbiome profile than those who were not taking the drug. This included an increase in microbes that produce short-chain fatty acids, which are known to decrease blood glucose levels.

We often don't take the gut microbes into consideration when we're thinking about a drug that works well on humans, but the concept of drugging the bugs is a new and exciting angle to potentially treat diseases.

12: Intestinal Diseases: Fire in the Gut!

The Gut: A Thirty-Foot Tube, but Mind the Gap

Nearly every discussion about the microbiota includes some aspect of the gut, including gut health and gut diseases. As we've seen, this is where incredible numbers of microbes happily live while we feed and water them every day and, for the most part, both we and the microbes seem happy with the arrangement. However, the intestines also provide an important barrier between the microbiota and your body, and sometimes there are problems with this barrier, resulting in nasty diseases. We've all heard of inflammatory bowel diseases (IBDs for short) such as Crohn's and ulcerative colitis, which are characterized by severe inflammation of the gut.

However, there are several other less obvious gut problems that the microbiota also seem to have a hand in. Some of these include colic (yes, think screaming, wailing infants and ultrastressed parents), celiac disease (gluten intolerance), and irritable bowel syndrome

(IBS). We're beginning to realize that the gut microbiota has an important role in all of these diseases.

You've probably never given it much thought, but the intestinal tract is an amazing organ that plays a critical role in our body. At a gross level (pun intended), we have a thirty-foot (nine-meter) tube running though our body that starts at the mouth, hooking up to the stomach, then the small and large intestines, and finally ending at the anus. Anything within that tube is actually not considered "inside" us, but transiting through us. The gut, including the intestine, has two main functions. The first is to form a barrier to keep everything we ingest inside the tube (food as well as all the microbes). However, its other function, which is contrary to its barrier function, is to digest and absorb nutrients and fluid from the intestine. Thus it has a tricky balancing act of being both permeable to things we want to take up, but impermeable to things that we don't. Luckily, nature has sorted this problem out, and a normal gut performs both of these functions very well.

Proper intestinal barrier function is critical for health. If spaces between cells widen, the permeability increases, and the general contents of the gut (including microbes and their molecules) can directly enter the body. This triggers a strong inflammatory reaction from the body, which is a common feature in IBD. Ironically, inflammation also seems to increase gut permeability, which can worsen the symptoms.

Similarly, as we will see when we discuss celiac disease, increased gut permeability presumably lets through more food particles, such as gluten, which causes the body to react to it. However, it seems that the presence of certain microbes tightens up the gaps between intestinal cells and decreases gut permeability. Studies show that the colonization of germ-free (GF) mice with bacteria early in life seals up their guts, while those that remain microbe-free have leaky guts—indicating that this is yet another function we rely on our microbes for.

To enhance its ability to absorb and transport fluids and nutrients, the gut has a very large surface area. Several reports suggest that if the intestine were flattened, it would cover an area nearly the size of a tennis court (2,800 sq. ft. or 260 m²)! A more recent study, using sophisticated microscopy and measuring techniques, showed that the human gut has a surface area of "only" 322 sq. ft. (30 m²), about the size of a studio apartment. Still, imagine cutting out a very thin cloth the size of an apartment and stuffing that into a tube the width of a sock. Anyone who has tried to roll up a tent and put it into its stuff sack will know exactly what we mean! The way the body achieves this remarkable feat is to have folds in the cloth (called *villi*), and then have many tiny fingerlike projections (called *microvilli*) on each fold. Think of a shag rug—each little projection on the rug is like a microvillus finger. You can see how this increases the surface area remarkably. However, remember that this entire area is exposed to microbes, and it cannot have any holes in it at all. Oh, and now also make each projection move in unison—that is how gut motility works. Finally, to move things along, the body coats the entire large intestine in mucus, a slimy substance made of proteins and sugars. Mucus also keeps many of the microbes at bay, at a distance from the microvilli. A thick mucus layer is associated with a healthy gut, and in intestinal diseases such as IBD or diarrhea it becomes thin. Many types of bacteria feed on mucus, and this actually helps maintain a thick mucus layer, as the body produces more in response to microbial mucus munching (mmm . . .).

As we've mentioned before, babies are not born with a fully functioning gut. In fact, quite the opposite: at birth babies are pretty much sterile, their gut is leaky, and their microvilli are not fully formed yet. Microbes kick-start many physiological processes in the gut, but the initial weeks of life are a period of adjustment while the gut settles into a more controlled environment. Recent studies point to the assembly of the initial microbial communities in a baby's gut as a factor that influences how a baby's intestines adjust to the first weeks of life.

For Crying Out Loud

We all know babies can cry quite a bit, and this noise certainly gets our attention (plus that of everyone else on the plane). It's been suggested that the reason a baby's cry is so noticeable is just that—she ensures she isn't ignored, since she can't speak up when her diaper needs changing or when she's hungry. However, some babies seem to cry nearly all the time, and this is called colic, something that can turn the early days of parenting into a complete nightmare for even the most doting parents.

Anamaria and Pedro were at their wits end. Their beautiful daughter, Sofia, was one month old and she cried day and night. Sofia would often cry so hard that she would turn blue. Their poor little baby would cry to the point of losing her voice until she eventually fell asleep, completely exhausted. Even while she was sleeping, Anamaria and Pedro would notice that Sofia was uncomfortable and fussy, and she would often wake up crying. During her whole second month of life, Anamaria recalls Sofia as either crying or sleeping; there were no cooing sounds, no baby grins or relaxed playtime.

To make matters worse, Sofia would flare up in rashes around her mouth and her chest. *This just can't be normal*, they thought.

As first-time parents, Anamaria and Pedro knew that taking care of an infant was supposed to be hard and that babies cry a lot, but Sofia acted like she was in pain all the time. Anamaria asked her mother for help and advice, but her mother just kept saying that Anamaria's brother was like that when he was a baby and that colicky babies simply need to be constantly held. Despite her mother's well-intended advice, they felt they needed medical help.

Throughout the following two months, they went to a total of five pediatricians to seek advice. The first three doctors dismissed their worries by saying that some babies just cry, and that Anamaria and Pedro were stressed because they were new parents and tired. Another doctor finally diagnosed Sofia with severe colic—a condition defined as excessive and inconsolable crying lasting three or more hours per day. He suggested that Anamaria's milk was causing Sofia's stomach pain, making Anamaria feel awful and guilty. He also said that she should stop breastfeeding and start using an extensively hydrolyzed (broken down) hypoallergenic formula, which they immediately did. The first time they tried the formula they could see an improvement in Sofia's colic, which gave them hope that the worst was over.

They followed up with a visit to a physician who specialized in pediatric gastroenterology. She tested a sample of Sofia's feces and when the results came back she determined that Sofia was allergic to a protein in cow's milk. The doctor was emphatic that breastfeeding should continue, but Anamaria had to follow a strict diet, avoiding dairy and soy products. Seriously committed to resume breastfeeding, Anamaria followed the diet and began to pump milk religiously. However, it had been a few weeks since she had stopped breastfeeding and despite all her efforts, she wasn't able to keep up with Sofia's feeding needs. As an alternative, the doctor suggested a different type of formula, to which Sofia did not react well at all. Her crying resumed and her rashes came back. After trying a couple more formulas, they found the right one for Sofia.

Anamaria and Pedro are not alone; one in five families deals with babies with severe colic, a condition that continues to increase in incidence around the world. Infant colic peaks at around six weeks and wanes after the infant is 3–4 months old. Although it lasts only a couple of months, it can be a devastating ordeal for the whole family and it has serious consequences, as severe colic can lead to parental emotional distress, anxiety, depression, and, on occasions, even child abuse. These families often require psychological therapy to deal with the stress related to colic.

In a sense, Anamaria and Pedro were lucky because they found the cause, but only about 5–10 percent of colic is a result of cow's milk allergies—the other 90–95 percent of cases have no known cause. There's no correlation with the sex of the infant, mode of delivery, breastfeeding, or birth weight. However, the microbiome still comes into the picture.

Studies of very young infants (in their first one hundred days) showed that infants with colic have a decreased diversity in their microbiota (remember, microbiota diversity is a good thing), plus a decrease in certain infant microbes acquired from the mother (*Lactobacilli* and *Bifidobacteria*) and an increase in bacteria that potentially produce gas and intestinal problems (*Proteobacteria*). They found that these differences start at one to two weeks of age, and that they were evident in all kids by one month of age, which is before colic sets in.

The types of bacteria that infants with colic lack are responsible for making substances that have anti-inflammatory effects and decrease pain (butyrate). In addition, colicky babies have more microbes that cause intestinal inflammation and gas production. The change in intestinal microbiota may also affect gut motor function, which could contribute to abdominal pain (and more crying). Thus, it's now being suggested that colic (at least those cases not caused by cow's milk allergies) could be thought of as a microbiota problem.

So, the million-dollar question to the suffering parents of children with colic is: If the problem is with the gut microbes, can we fix it? Maybe. Two recent small studies using a probiotic (*Lactobacillus reuteri*) found that it decreased crying by twofold (heck, one and a half hours of crying versus three hours? Sign us up!). However, another recent study used the same probiotic and found it had no effect. Clearly the jury is still out regarding the treatment of colic with probiotics.

There are two main messages we can take home here: 1) the microbiota is different in kids with colic, and this change is detectable right after birth, even before colic sets in; and 2) altering the microbiota, if done right (and early in life) might just improve it. As with so many of the topics in this book, we're just realizing that the microbiota play a pivotal role, and we still don't know the exact good and bad bugs involved, or the perfect combination of bugs to fix it. The fact that studies are under way to test the administration of probiotics to prevent colic is of little consolation to those parents who have an infant with colic right now, but hopefully we'll know more soon—or the kid will grow out of it.

Chewing on Gluten:
Microbes and Celiac Disease

Gluten is a natural complex protein (actually two proteins together) that is found in wheat, rye, barley, and other grain products. It gives bread its chewiness and helps bind baked goods together. The next time you're at a pizza house watching the staff toss a pizza crust high in the air and stretch it, remember it's the gluten that's holding it all together. However, wheat gluten has increasingly been associated with immune reactions against it, including celiac disease (barley and rye have similar proteins that can also cause this disease). Between 1–3 percent of the population suffers from celiac, and this disease has increased more than fourfold in the last fifty years, lumping it in with the ever-growing incidence of "western lifestyle" diseases. It requires a human genetic component (which means you have to have the right genes to potentially get it), as well as environmental factors (which include the microbiome).

About one-third of the human population carries the genes that are a prerequisite for celiac (called HLA-DQ2 and HLA-DQ8). Having these genes means you're at risk for the disease, but it certainly doesn't mean you'll get it. It's a disease of the small intestine, where the presence of gluten triggers a strong immune response, causing gut inflammation and damage to the surface of the small intestine. The symptoms are miserable, and include diarrhea, abdominal pain, and weight loss. They can also be subtle, causing stunted growth in kids and iron deficiency. The most obvious way to handle this disease is to avoid foods containing gluten, also known as a gluten-free diet. Such diets are usually quite effective at treating the symptoms, but not always. Because wheat is a moderately recent addition to the human diet (we figured out how to cultivate it about ten thousand years ago), gluten-free diets are sometimes considered to be more like the diets we evolved on (scavenging nuts, seeds, etc.), and are thus becoming popular in some circles, even for those that don't have celiac disease.

Several smoking guns implicate the microbiota in this disease, but unfortunately, we don't have the full picture yet. A major risk factor for celiac disease in children (assuming they have the right genes) is an infection or being treated with antibiotics, especially penicillins and cephalosporins, in the first year of life. However, taking antibiotics during pregnancy did not seem to affect the risk of the child developing the disease.

Having an elective C-section also increases the risk, but an emergency C-section does not, which at first seems weird. However, many emergency C-sections are done in the second half of labor, once the baby is on its way down the birth canal and after the membranes ("water") are broken. It's thought that during such C-sections, the infant would have already been in the birth canal and thus was exposed to the maternal microbiota. Breastfeeding is also recognized to decrease celiac disease.

When the intestinal microbiota is studied in people with celiac disease, differences are found when compared to people without the disease, but the results vary widely between studies and there's no consensus regarding the composition of a celiac microbiota other than that it's different. Then again, celiac sufferers have an inflamed gut that would contain different microbiota anyway. Also, celiac disease is a disease of the small intestine, and fecal samples (from which these studies are done) are more indicative of large intestinal microbiota (the small intestine is much harder to sample). One trend that does come through is that the microbiota changes are generally similar to those described above for colic, with decreases in the good bacteria and increases in the inflammatory ones.

Putting people on gluten-free diets often, but not always, restored their microbiota to a more normal type, but again this could simply be due to the decrease in gut inflammation. Similarly, people on gluten-free diets who still show symptoms have altered microbiota, as well as gut inflammation.

What might the microbiota be doing? Again, there's little hard data on this, and much speculation. The microbiota may be generating products that affect the immune system's response to gluten, especially early in life. They may be making toxic products that increase intestinal permeability, which could lead to an increase in gluten penetrating into the body, triggering an immune response to it, or influencing the immune system in some other way that affects its tolerance to gluten.

The good news is that there's excellent data coming out of studies defining when gluten should be introduced into a child's diet. Studies done in the 1970s showed that introducing gluten and solid foods into a diet before four months of age increased the occurrence of celiac disease, possibly by exposing the child to gluten too early (into a leaky gut that is still developing) and triggering an intolerance. Other studies showed that introducing gluten at seven months of age or older also increased the risk of disease, possibly because these kids can eat more and may be getting a large dose of gluten the first time it's introduced. It appears that the sweet spot for gluten introduction is between four and seven months of age. It's important to add gluten to the diet in small amounts *and* to continue breastfeeding. Breastfeeding may introduce small amounts of maternal gluten or gluten antibodies to the child, decrease early infection rates, or just affect the microbiota composition (nobody really knows).

What about probiotics? In animal models of celiac disease, probiotics worked by decreasing inflammation and the disease. However, data for humans are very scarce, with only one study; it showed that probiotics had a slightly beneficial effect, but it wasn't statistically significant. As the role of the microbiota is further understood for this disease, it's likely that more targeted probiotics will be developed, which will include microbes that specifically digest and break down gluten.

Irritable Bowel Syndrome

By far the most commonly diagnosed gut problem is irritable bowel syndrome (IBS), affecting up to one in five people, often teenagers. It's not a single disease, but rather a set of symptoms that, as the name aptly implies, irritate one's bowels, or gut. The symptoms vary from diarrhea to constipation (or both), bloating, excess gas, and abdominal pain —all of which makes a person very uncomfortable. Unlike the other diseases in this chapter, there don't appear to be structural changes to the intestine with IBS. Stress, anxiety, and depression are often associated with it, hinting at links between the gut and the brain (see chapter 14).

Stress is the most commonly acknowledged risk factor of IBS (we know that stress affects the microbiota based on a study of university students writing final exams). Antidepressants are commonly used in moderate to severe cases of IBS, helping with pain perception, mood, and gut motility.

There are strongly established correlations between the microbiota and IBS. About one-quarter of new IBS cases happen after an intestinal infection, which we know impacts the gut microbiota. Antibiotics and diet changes, which alter the microbiota, can also trigger IBS, and patients often claim a particular food seems to be a trigger. Probably the most compelling evidence comes from experiments in GF animals. When human feces from IBS patients were transplanted into microbe-free mice, they developed IBS symptoms. However, when feces from healthy individuals were transplanted, no IBS symptoms were observed. There are certainly differences in the microbiota composition in IBS patients, although no defined "microbial signature" has been identified yet. IBS patients have a decreased diversity in their microbiota, and bacteria normally found in the large intestine are often found in the small intestine in IBS patients, reflecting a microbiota imbalance.

Certain clues suggest that if one alters the microbiota, symptoms of IBS may decrease. For example, treatment with an antibiotic that remains in the gut and is poorly absorbed (rifaximin) decreased intestinal symptoms, although they did redevelop later. Similarly, there are reports of using prebiotics and probiotics with some level of success in treating symptoms, but because of the small trial size and varying probiotics, currently no general recommendations can be made for probiotics and IBS. There has been one small fecal transfer trial of thirteen people, and it resolved symptoms in 70 percent of the patients—this was after dietary modifications, antibiotics, probiotics, and/or antidepressants had all failed! This is particularly good news for a miserable and tough-to-treat disease, and begs further trials in this area.

Another treatment that shows much promise is a particular diet, low in FODMAPs (fermentable oligosaccharides, disaccharides, monosaccharides, and polyols—jargon for foods that contain certain sugars and sugar alcohols). These small compounds are poorly absorbed in the small intestine and accumulate in the gut, causing water to increase in the intestine, which leads to diarrhea. However, they are readily digested by gut bacteria, producing gases that can also contribute to IBS symptoms, such as bloating and pain. Studies in Australia have shown that this diet reduced IBS symptoms, and it is now being recommended to treat IBS. A low-FODMAP diet is not a DIY diet, and needs to be undertaken with a dietician specially trained in this area. In the beginning, all high-FODMAP foods are omitted from a person's diet. It takes 6–8 weeks before certain foods are reintroduced, and eventually only a small number of foods are usually excluded. It should also be noted that some FODMAPs are needed for our body (and our microbes) to function properly, and it is *not* a zero-FODMAP diet, just one that has fewer of these components.

Inflammatory Bowel Diseases

The last of the major gut diseases we'll discuss are the inflammatory bowel diseases (IBDs), which include two major, related intestinal diseases: Crohn's and ulcerative colitis. As the name IBD implies, they feature inflamed intestines that don't resolve, making patients' lives miserable with persistent diarrhea, rectal bleeding, abdominal cramps, pain, and weight loss. About 1 in 150 individuals suffer from these diseases in Canada and 1 in 300 in the US. Attempts have been made to control the gut inflammation with anti-inflammatory drugs, but these have limited success, and about 75 percent of Crohn's patients ultimately need bowel surgery to remove damaged and destroyed portions of their intestine.

The onset of these diseases is usually in young adults (ages 15–30), although children can also have them. The rate of IBD in the US and Canada has plateaued, but it is increasing rapidly worldwide as other countries become more developed and adopt a Western lifestyle.

Like most gut diseases, both host genetics and the environment (including the microbiome) play a major role. Scientists have identified 163 human gene mutations that increase the risk of disease, but no single gene has been identified as causing it. Instead, there's a collection of risk factors, both human and environmental. The genetic mutations often occur in biochemical processes associated with inflammation. It's thought that these genes play a role in controlling the gut microbiota and, if mutated, they're less able to contain the microbiota within the intestine. Any loss in gut barrier allows microbes to penetrate through, which then triggers extensive inflammation as the body tries to repel the microbial invaders.

A major problem with IBD is the need to go to the bathroom frequently (up to twenty times per day!). This is tough if you're travelling, or out of your regular neighborhood. Crohn's and Colitis Canada has started a program called Go Here, placing signs in windows of establishments where you can use a washroom immediately, no questions asked. The best part is that there's an app for it, so a smartphone can rapidly identify the location of the nearest washroom.

The microbiota is implicated in IBD, but once again no causative microbiota have been definitively established. We know that GF mice do not get IBD, presumably because there are no microbes to breach the barrier and trigger inflammation, even though GF mice have a leaky gut. However, when particular microbes are introduced into GF mice, they do cause varying degrees of IBD.

As expected, the gut microbiota is altered in IBD patients at the time of diagnosis, before treatment has been started. It's known that antibiotic usage can trigger IBD, presumably by altering the microbiota. As we've seen with the other diseases in this chapter, as microbiota diversity is threatened, the number of inflammatory microbes increases, and the number of beneficial, anti-inflammatory microbes decreases.

There has been a major effort to control and treat IBD diseases by implementing fecal transfers and recommending specific diets. Although fecal transfers have been used since 1983 to treat *C. diff* disease, with great success (at a >90 percent cure rate; see chapter 16 for more on this), only recently have they become extensively used for IBD, and the results have been mixed. However, this makes sense because, with *C. diff*, it doesn't really matter which microbes are used or who the donor is—as long as some microbes are added, they will displace *C. diff*. However, in IBD, the person already has an inflamed gut, so these transfers are like putting microbes into a fire and asking them to survive and put it out. It appears to matter which donor is used for a successful outcome. When we look collectively at the several small trials that have been done for IBD, they indicate that fecal transfers work better for Crohn's disease than colitis, and better in pediatric populations than with adults. Together, they have about a 45 percent clinical remission rate, which is promising, but still needs more work (see The Scoop on Poop Transfers, page 187).

Finally, similar to IBS, the low-FODMAP diet is being tried for those with IBD, and in one small trial, 50 percent of patients saw a marked reduction in their symptoms (although this doesn't treat their disease, just their symptoms). There are several more low-FODMAP diet trials under way, so we should know fairly soon how effective it really is at improving symptoms.

Dos and Don'ts

- ◆ **Do—** ask your doctor if a cow's milk allergy might be responsible for your child's colic. This is the one cause of colic that we know can be treated (by removing cow's milk from the mom's diet). Also discuss a course of probiotics for your child with your doctor, as these are safe, and they may significantly improve the condition. And remember: the silver lining of colic is that it spontaneously goes away after a few months of wailing—the trick is to survive those few months.

- ◆ **Do—** introduce small amounts of gluten into a child's diet between 4–7 months of age, while continuing to breastfeed; avoid introducing gluten before four months or in large quantities for the first time after seven months. By continuing to breastfeed while adding solid foods, you are decreasing your child's chances of developing celiac, as well as continuing to improve your child's microbiota.

- ◆ **Don't—** try to tackle a low-FODMAP diet yourself, as it's quite complex and requires the supervision of a trained dietician. However, talk to your gastroenterologist about whether a low-FODMAP diet might give you results, and get a referral for the right professional. Such diets are known to improve symptoms in a significant number of patients.

- ◆ **Don't—** attempt a fecal transfer yourself (do we really have to say this?), but consult with your physician regarding whether this might be an option for you, if you suffer from IBD or IBS.

THE SCOOP ON POOP TRANSFERS

Fecal transfers for IBD have a nearly 50 percent success rate over-all, which is great . . . but can it be improved? Just recently two major clinical trials of fecal transfers for ulcerative colitis were published. Although one study didn't see any beneficial effect, the other one was quite interesting: it wasn't working until Donor B, who apparently had the right fecal microbial mix, showed up and then the trial was a success. Hopefully, they're assessing Donor B's microbiota (as well as those from donors that didn't work) in order to define the characteristics of a good donor. Five more major clinical trials are under way, so we'll soon have a much better idea about the use of fecal transfers to treat IBD.

There are several ways one could potentially improve the odds of a successful fecal transfer, and charitable groups that are dedicated to improving treatments for IBD, such as the Kenneth Rainin Founda-tion, are looking closely at the various options. These include using donors like Donor B (we would love to know what they eat and drink!), who are known to "have the right stuff" (although we don't know exactly what that is yet).

Another possibility is to decrease gut inflammation with anti-inflammatory medication before the transfer is given. Because IBD, by definition, means inflammation in the gut, this might provide the incoming microbes a fighting chance to successfully colonize, and perhaps displace the previous microbes. Similarly, putting patients on a favorable diet could alter the microbiota and decrease inflamma-tion prior to a fecal transfer. One might consider antibiotic treatment to get rid of the "bad" bugs and clear out the intestine prior to trans-plantation to allow the incoming bugs the opportunity to colonize. Or one could use repeated fecal transfers to try to increase the odds of seeing a beneficial effect.

Unfortunately, like most things in medicine, to try all these permutations and combinations in a major clinical setting will take many hospitals, numerous patients, a lot of money, and multiple years to complete before the ideal fecal transfer setting is established. However, given the major impact of this disease, we're optimistic that such efforts will succeed.

13: Asthma and Allergies: Microbes Keep Us Breathing Easy

The Burden of Asthma

When she was a child, Claire would wake up in the middle of the night to the sound of her sister's wheeze far too often. The forced, fast-paced whistling raised her sister's upper body with every exhalation, seemingly taking all her energy with every breath. This sister, Stephanie, was only ten years old (Claire was eight), but had dealt with sleepless nights like this all her life. Their mother would usually give her medicine before going to bed, propping Stephanie up with two or three pillows and singing to her until the child fell asleep.

Even at that young age, Stephanie would practice different techniques before going to bed to prevent her asthma attack from getting out of hand; she would try to remain calm (hard to do when you feel like you're asphyxiating), avoid gasping, and, above all, control the need to cry or cough, as these only made things worse. Yet sometimes her breathing would become unmanageable and she would ask for help.

Claire shared a room with her sister for as long as she could remember, and she could tell when Stephanie couldn't deal with the asthma attack anymore and it was time to wake their parents. They had bought Stephanie a personal nebulizer, which probably cost an arm and a leg, but was probably the best purchase they ever made and was a game changer for their family (home nebulizers had just become available in the mid-1980s). Before owning a nebulizer, a night like this one would invariably result in a trip to the ER that would either last until the morning or result in a multiday hospitalization.

Claire remembers the night her mother asked her to prepare Stephanie's nebulizer by herself for the first time. It involved attaching a syringe to a needle, drawing up a small amount of Ventolin (a common asthma medication), another amount of saline solution, and combining them in a small cup. This little cup would then get plugged into an air compressor at one end and into a mask on the other end, which Stephanie would strap to her face.

To this day, Claire remembers how the noise from the compressor sounded and how the sweet-smelling steam from the medication would fill up the room within minutes, allowing her sister's breathing to become a bit deeper and slower. Little by little, her breathing would relax until everyone fell asleep until the morning.

For Stephanie, asthma season lasted from about August to December, when the rain was the heaviest and the humidity in their native Costa Rica caused all sorts of allergens to trigger her asthma. Her attacks could be extremely severe and land her in the hospital for days or weeks, making their mom refer to the local children's hospital as their second home. On two occasions her asthma got so bad that she went into respiratory and cardiac arrest and was brought back to life by paramedics. It was a frightening time for Claire's entire family.

Fortunately, asthma treatment has improved over the years, and now Stephanie, almost forty, keeps her asthma in check with a medicine cabinet full of inhalers and pills. But about once every year she can't escape a bad cold that spirals downward into another asthma crisis. Her lungs have accumulated too much damage over the years, leaving her with less than half of normal lung capacity. Undeniably, asthma is a terrible disease that takes a toll on a person's physical, emotional, social, and academic life, and on their family as well. Imagine if there was a way to prevent asthma from developing in the first place.

Too many families have had to deal with the consequences of having an asthmatic child. This chronic disease of the lungs is characterized by inflammation of the airways and a sudden narrowing of the bronchioles, the smallest branches of the airways, which then makes breathing difficult. It's thought that asthma develops due to a combination of genetic and environmental factors, an explanation that scientists commonly use when there are just too many things involved in a disease and we don't yet know how it all works.

What is known is that the rates of asthma have skyrocketed in certain parts of the world. A case as severe as Claire's sister's was considered rare back in the 1980s, but asthma has been increasing in incidence for the past three decades, and so has its severity. In just one generation, rates of asthma have tripled, with 10–20 percent of children currently suffering from asthma in North America, Australia, and most Western European countries, affecting an estimated 300 million children worldwide. It's the most prevalent chronic pediatric disease in the world, the number one cause for hospitalizations of children and missed days of school. In contrast, the rates of asthma in underdeveloped countries did not change much during that period.

Today asthma cannot be cured. It's also difficult to treat, and once treated is even more difficult to control and prevent from flaring up again, making it one of the most expensive diseases for public health systems around the world. Even more worrying, asthma rates in highly populous, less developed countries are now beginning to rise. All this makes us wonder: What is going to happen two generations from now, if asthma continues to increase at these rates? Are asthma inhalers in the future for every single one of our grandchildren or great-grandchildren? Is this disease going to become part of the human condition just like dental cavities?

Searching for the Culprits

The fact that this disease, along with the other Western lifestyle diseases discussed in this book, has increased so sharply in such little time is a true enigma. There are known genetic predispositions for asthma: for example, in Claire's family asthma can be traced back four generations. Genome-wide association studies (also called GWAS, a name given to large studies that look for gene and DNA variation in people with and without certain diseases) found several genes associated with asthma, but they don't explain why the majority of asthma cases are inherited. More importantly, these genetic alterations fail to explain why asthma has become such an endemic disease. Our genetic makeup simply has not changed that much in just one generation—it has to be due to something else that has changed in the environment.

Research groups have looked at things such as diet, socioeconomic status, sun exposure, pollution, pollen, ethnicity, contact with animals, urban vs. rural environments, exposure to specific insects, and more. Many of these studies have yielded strong associations between a particular environmental factor and the risk of developing asthma, but the strongest and most consistent one is the farm environment. People who grow up on farms have a much lower risk of developing asthma than anyone else in Western societies. Something in their lifestyles has protected them from the dramatic increase in asthma, but what is it?

In trying to figure this out, one of the most studied groups are the North American Amish, who have the lowest reported prevalence of asthma and allergies of any Western population. Dr. Mark Holbreich, an allergist from Indianapolis, noticed the low rates in allergies in Amish communities in Northern Indiana after treating them for over twenty years. He is the lead author of a study published in 2012 in which children from 157 Amish families were compared to 3,000 Swiss farming families and 11,000 Swiss nonfarming families. They found that only 5 percent of the Amish children had asthma, compared to 6.8 percent of Swiss farm children and 11 percent of Swiss nonfarm children. The incidence of asthma in Amish children is comparable to what it was for everyone else a few generations ago, making them an ideal population to examine further.

The Amish and their traditional lives, without cars, electricity, or modern appliances, feels like going back to the mid-1800s. Entire families work the land and tend to farm animals, and live a technology-free life. Children as young as five years old milk cows, three-month-old babies visit the cowsheds, and many of them even learn to walk there. Contact with farm animals and dirt clearly occurs very early in life. Pregnant mothers work in the cowshed throughout their pregnancy, potentially exposing their babies to microbes prenatally as well.

However, not all farms are created equal in terms of asthma protection. An interesting epidemiological study that looked at the rate of asthma and allergies in Hutterite communities found that they do not appear to be as protected from the increasing asthma incidence as the Amish people are. Hutterites share certain similarities with the Amish; they both live communal, self-sufficient lives based on strict religious beliefs, and they're both of German descent. However, Hutterites have an important distinction: unlike the Amish, they welcome technological advances, use state-of-the-art farming equipment, and usually run quite large farms. Hutterites also use antibiotics in their farm animals in a way similar to most other modern farmers.

Studies of European farms, many of them led by Dr. Erika von Mutius at Munich University in Germany, have shown that farms with a higher diversity of microbes and in which the microbes from the cowshed reached the home offered the strongest protection from asthma. In one study, the amount of microbes in the mother's mattress was inversely correlated with the risk of her children developing eczema, an allergic skin condition often associated with asthma. All these studies clearly suggest that, when the farm's microbial residents come in contact with the human residents early in life, children are somehow protected from developing this disease.

A separate study, also led by Dr. von Mutius's group, found that newborn babies from farms are born with an immune advantage over those not from farms. Researchers isolated immune cells from the umbilical cord blood and found that newborns whose mothers lived on farms during their pregnancy had an increased number of regulatory T cells, which are important immune cells that serve as the peacekeepers and modulators of the immune system (see chapter 2), making sure that the immune system does not overreact towards a particular intruder. They're also known to be crucial in preventing asthma and allergies. Since the researchers compared groups that share the same ethnicity, the differences they found are not due to genetic variations, but likely due to differences in environment. The thought is that farm exposure in pregnant mothers reflects a more natural way of fetal immune development, one that involves the stimulation of a particular group of cells necessary for immune control and for prevention of allergies. By being in contact with cows, chickens, pigs, and dirt, mothers are essentially immunizing their children to tolerate future allergens (substances that cause allergies) by more closely mimicking how humans evolved as a species, in close contact with animals and the natural environment. Removing this exposure during pregnancy, as happens in most modern Western environments, is likely one of the drivers of the sharp increase in asthma risk in everyone but farmers and their families.

From the Gut to the Lung

All these studies about the patterns of asthma incidence made it clear to us that, in order to figure out what the protective factor is and how it works, scientists needed to focus on microbial exposure and how it interacts with the immune system. An idea on how to do this came to Brett while he was having dinner with his wife, Jane. She has worked as a pediatrician for thirty years and she told him that there was some evidence that children who receive antibiotics early in life have a higher risk of developing asthma. Brett thought this concept was fascinating, but was skeptical. But as always, Jane was right—the literature backed her up.

Brett became very interested in this and brought the discussion to his lab. At that time, they were using antibiotics as a tool to shift microbiota and were then looking at the effect of these shifts on diarrheal infections in the gut. Brett talked to one of his students, Shannon, and she decided to tackle a very important question during her PhD research: Do changes in intestinal microbiota affect susceptibility to asthma? This was only five years ago, but back then it was considered a far-fetched idea and she even had some side projects on the back burner, in case this didn't work.

Shannon decided to do a simple experiment. She set up two groups of mice: one group would receive antibiotics starting from birth until they became adults and the other group would receive antibiotics starting at around three weeks of age, when mice are weaned and not considered babies anymore (there was also a control group that didn't receive any antibiotics). When Shannon tested the mice for asthma, the results surprised everyone. The mice that received antibiotics when they were babies had worse asthma than the mice that received them only in adulthood or those that didn't receive them at all. Furthermore, she analyzed the type of immune cells in these mice and found that, similar to what had been seen in humans living on farms, the mice that received antibiotics during infancy had lower amounts of regulatory T cells (the peacekeepers) in the intestine. The antibiotic that Shannon used was given orally and does not get absorbed into the bloodstream and the rest of the body, which means that somehow the shift in microbes in the intestine changed the outcome of an immune disease that occurs in the lungs. Her finding that regulatory T cells are involved also suggests that what she observed in mice probably has a connection with what occurs in humans. Shannon went on to do many more experiments, and she showed that the effect of antibiotics on asthma was limited to very early in life, from right after the mouse pup was born until it was weaned, somewhat equivalent to the first few months of a human infant.

Claire joined the lab at this point, and she was eager to carry on with this project using human samples. Our lab was fortunate to form a partnership with a national study called the Canadian Healthy Infant Longitudinal Development (CHILD) Study, which had been looking at factors that might affect asthma in children. The CHILD Study had been collecting samples across Canada since 2009 and its goal was to collect and characterize samples from 3,500 children from birth until five years of age (a massive effort). Claire and her collaborators requested fecal samples from 350 children, which were collected both at three months and one year of age.

Our question was simple: Are there intestinal microbial differences in babies that go on to develop asthma earlier in life compared to babies that do not develop asthma? Using a new method to survey bacteria (16S analysis; see chapter 16), we sequenced the intestinal microbiota in these samples and found something that really surprised us. Three-month-old babies at a high risk of developing asthma later in life were missing four types of bacteria, but when we looked at the microbiota at one year of age, the differences were essentially gone. All of this—plus the other studies about antibiotics, farms, etc.—points to a critical window of time in which microbial changes in the intestine have long-term immune consequences in the lung. In addition, we found that the differences were not limited to the type of bacteria found in feces but were also observed in some of the compounds they produce. Interestingly, only one of these compounds was a bacterial product made in the gut (acetate; see chapter 2), and many of them were bacterial compounds detected in the urine of these babies— another proof that bacterial metabolites go everywhere in our body.

Our lab is still trying to figure out how these four bacteria (which are present in low numbers and which we nicknamed FLVR) lead to asthma, but one additional set of experiments in mice suggests that FLVR is directly involved with mediating this, as opposed to just showing a correlation. Claire gave one group of GF mice a fecal transplant from a baby who had no FLVR in his feces (and also developed asthma later in life), whereas a second group of mice received the same fecal transplant plus added FLVR. When the mice became adults, the group that received FLVR had much less lung inflammation and other markers of asthma.

We're still far from considering this as a preventative therapy for human asthma, but it opens a few doors that might give us a major advantage in combatting the asthma epidemic. The first is that, assuming our findings hold true in other populations, we should be able to identify infants that are at very high risk for developing asthma. Even more exciting, we may be able to give those high-risk infants certain microbes or microbial products as a way to prevent asthma. It really is a bizarre and almost unorthodox concept to think that changes in intestinal microbiota in a three-month-old could affect an allergic lung disease in a school-age child several years later. Stay tuned!

Allergies and Eczema, Too?

When Claire's daughter, Marisol, was two months old, her skin would break out in red glossy patches and she would cry while constantly trying to rub her skin. By the time she was six months old the patches were in almost every one of her chubby body folds. Her scratching got so bad that she would break her fragile skin and Claire would sometimes find blood on her crib sheets. Claire tried different creams and lotions recommended by their pediatrician, but when her daughter's rash was severe, the only treatments that would control it were topical corticosteroids (which decrease inflammation) and fewer baths.

During the dry winter months, Claire would bathe her daughter only once a week, moisturize daily, and apply anti-inflammatory creams if her skin flared up.

When Marisol turned one she experienced a reaction the second time she was offered eggs. Claire and her husband didn't introduce allergenic foods until then, as they knew that Marisol was an allergy-prone child, given her early start with eczema (those red patches described above). They realized that she probably had a higher chance of developing food allergies and/or asthma later on, so they needed to be careful. The statistics were not in Marisol's favor; she had a family history of allergy and asthma on both sides—she's what is called an atopic child. Atopy is a predisposition towards developing allergic diseases, driven by an overly reactive immune system. Luckily, Marisol's eczema is kept under control fairly easily these days (she's five now); she completely outgrew her egg allergy; and, most importantly, she hasn't developed asthma.

At the heart of all this allergic disease is an unbalanced immune system. The immune system can affect different organ systems, causing eczema, asthma, and hay fever. This is why understanding the microbiome is so important. The microbiome has the potential to train the immune system and prevent all forms of allergic disease.

Atopic children can manifest skin, respiratory, or food allergies, separately or in any combination. More often than not, the appearance of these conditions follows an order known as the atopic march. This "march" often starts with eczema (also known as atopic dermatitis), followed by food allergies, allergic rhinitis (hay fever), and finally asthma. All of these diseases have been increasing in incidence, with eczema affecting approximately 20 percent of children in wealthier countries of the world; many of these children go on to develop asthma. It's been shown that the likelihood of a child with eczema to develop asthma is associated with how severe the skin condition is, making efforts to control or treat eczema an important step in preventing worse allergic conditions from developing.

Unfortunately, eczema doesn't have a cure, but it can be effectively treated using a combination of bathing techniques, avoidance of skin irritants, and anti-inflammatory treatment (such as corticosteroid cream). However, recent data show that the early gut microbiota of children that later develop eczema is different and less diverse than that of control children. In addition, the microbiota of affected skin areas is different than that of healthy skin. Researchers still don't know whether the difference in skin microbiota is a cause or an effect of eczema (the inflamed skin may alter the microbes), but it opens the possibility of treating the microbiota as a way to control eczema. Experimental approaches to treat eczema by treating the skin microbiota look promising and a few products are already in the market, but additional research is needed to demonstrate that it is indeed effective.

Some studies have used oral probiotics to prevent or treat eczema, but the results have been mixed. The strongest evidence shows that probiotic treatment during pregnancy and early infancy is effective at decreasing eczema risk (even in cases with strong family histories), but the evidence for probiotics as an actual treatment is weaker. Still, this area of research is extremely new, and the reason many of these treatments appear to be ineffective may be because current probiotics contain bacterial or yeast species that are not involved in the development of eczema or related allergic diseases. Our research on asthmatic children shows that the bacteria involved in protecting against asthma are not found in any probiotic formulation (yet). Needless to say, our laboratory and others are working hard to determine which microbial species should be targeted.

Another promising finding comes from a large study of 5,000

children, in which it was discovered that babies who were deficient in vitamin D were significantly more likely to develop food and respiratory allergies. Vitamin D is a known anti-inflammatory that suppresses the action of immune cells. Emerging evidence shows that vitamin D is also necessary for some of the cross talk that occurs between our gut bacteria and our immune cells.

Dos and Don'ts

◆ **Do—** take measures early on to prevent the development of asthma and allergies in your child, even before she's born! Maintain a healthy diet, avoiding tobacco smoke and anti-biotics if possible. Given the chance, opt for a vaginal birth, and prolong breastfeeding.

◆ **Do—** give your baby and toddler a vitamin D drops supple-ment daily. Deficiency of vitamin D has been shown to increase the risk of food and environmental allergies and is also a crucial component in regulating the microbiota.

◆ **Don't—** rush out to the nearest farm if you're a city dweller and are pregnant or have a young child and bed down with the cows. While farm families should be considered for-tunate to live in such an environment, some studies show that occasional visits to farms may actually exacerbate any pre-existing allergy tendencies. It seems that prolonged continual exposure to this environment early in life is what provides protection from asthma and allergies. But if you have a dog, snuggle up with it frequently—and let your child do the same, to bring some of the outdoors into your family (see chapter 8).

♦ **Do—** stay on top of literature on this topic. It's clear that the gut (and possibly the skin) microbiota is directly involved in asthma and allergy development, but it's still not known how microbial alterations can be used as a way to prevent these diseases.

ALLERGENIC FOODS: AVOID OR FIGHT BACK?

The common advice to parents of children with food allergies is: avoid that food! However, a relatively new treatment option is to face a food allergy head-on, under careful medical surveillance, until a child's body no longer reacts as fiercely to the food.

This therapy is known as allergy desensitization and is being offered experimentally in a handful of hospitals and doctor's offices. The treatment consists of first finding out the amount of allergen (e.g., peanuts, milk, etc.) that will provoke a reaction, and then beginning to feed the child increasing amounts of the food every week over the course of a few months. The goal of this therapy is not necessarily to cure the allergy, but to reduce the risk of a severe anaphylactic reaction if the child eats a small amount of the allergic food.

Allergy desensitization is quite effective, but upon "graduation" it needs to be maintained. Children that have been successfully desensitized must eat a small amount of the allergenic food every day (e.g., a few peanuts or a few sips of milk) to keep the allergy from returning.

This is a promising but potentially dangerous approach to treat a severe allergy, so it should be attempted only under the care of an allergist trained in this therapy.

14: Gut Feelings: Microbiota and the Brain

Bottom-Up Thinking

So far we've seen that changes in a developing microbiota can alter our lungs, skin, intestines, pancreas, liver, and fat tissue. What about the organ that controls our senses, intelligence, and behavior? Can microbes in our gut affect brain development and be involved in neurological disease? Well, sure.

We've known for a long time that there's a link between the gut and the brain. The gut has the second highest number of neurons (millions of them)—next to the brain, of course. These neurons combine to connect the entire digestive system, from the stomach all the way down to the anus, reporting back to the brain on how our digestive system is doing, and affecting gut movement and other digestive functions. An important nerve, called the vagus nerve, serves as a direct neural connection between the brain and the nerves in the gut; the vagus nerve links gut sensations to the brain. Remember the last time you peered over a steep cliff and felt that funny sensation in your stomach? Or when you had butterflies because you had to speak in front of people? What does your gut have to do with your physical responses to any of these situations? A tingling tummy won't help you avoid falling off a cliff or make your speech any better, but it does prove that your gut and your brain are in sync.

Until recently, this connection (called the gut–brain axis) was thought of in terms of a "top-down" concept, with the brain doing all the talking and the gut following orders. For example, in irritable bowel syndrome (IBS), it's thought that the brain contributes to the pain and other bowel symptoms, as there's often no known direct cause within the gut, and because things like stress affect it. But that's all changing, and we now realize that the gut, and especially the microbes in the gut, have a lot to say; a "bottom-up" approach is beginning to influence how we think about the brain and its functions.

Nature is full of amazing examples of how microbial pathogens can affect their host's behavior, with the microbe driving its host to do things that actually help the microbe (but not necessarily the host). We've all heard of the frothing, snapping rabid dog that tries to bite anything that moves. Rabies is caused by a virus that's transmitted through bites that inject the virus into a new host, where it enters the nerves and climbs up into the brain. Unless treated early, rabies is fatal. So how does a rabies virus find a new brain before its current host dies and is buried deep in the dirt? By affecting the brain, convincing the animal to bite as many potential hosts as possible. In humans, rabies has also been linked to aggressive behavior and, strangely, to hydrophobia, which is an intense fear of drinking water—sufferers cannot even bring a cup of water to their mouth (watch the YouTube videos for a scary demonstration).

Another fascinating example of behavioral change is carried

out by a parasite called *Toxoplasma gondii* that lives in the brains of mammals. *Toxoplasma* likes to live in cats and is fairly common around the world. It undergoes sexual reproduction in the guts of cats, which is key to its life cycle. *Toxoplasma* also lives in the brains of rodents such as mice and rats. So how does it get from a mouse brain to a cat gut? When this parasite is in a rodent brain, it somehow reprograms that brain to do two things: it decreases fear so that the rodent leaves its hiding places and wanders out in broad daylight, and also it attracts the rodent to cat odors! Humans are not the preferred host for *Toxoplasma*, but infection does occur (this is the parasite that pregnant women aim to avoid when instructed by their doctor not to change the cat litter box). Humans who have been infected with this parasite have a higher risk of suffering behavioral changes and schizophrenia.

There are many more examples of behavioral modifications by microbes, including convincing fish to seek out warmer water (to let the parasite grow faster), persuading grasshoppers to leap more and jump into water (for the parasite to mate and lay its eggs), and there's even a fungus that turns ants into zombies so they wander around aimlessly and then chomp on to a leaf and die, holding the leaf in a death grip (which the fungus then uses as a food source). These are pathogenic microbes commandeering their host. So what about our resident microbes—are they chatting with the brain, and could this affect our development, our behavior, and even have an impact on diseases of the nervous system? To a certain extent, yes. The microbiota can affect anxiety, depression, social recognition, and stress responses; it may even play a role in common mental disorders such as autism.

The Microbes Made Me Do It!

So how can a microbe that lives in the gut talk to the brain? As mentioned above, our nerves are hooked up to all parts of the digestive tract, which then feed into the vagus nerve that not only brings signals down from the brain to the gut, but also feeds gut information back up to the brain (again, the gut–brain axis). Any of these nerves can be used to send a message from the gut to the brain, using compounds known as neurotransmitters. Alternatively, microbes can either make molecules, or alter existing molecules, and send them as messages directly to the brain using the bloodstream. The third way is for microbes to tweak the immune system (something they're very good at), which is directly hooked up to the nervous system, and send messages this way.

Neurotransmitters are chemicals that send signals between two nerves, passing along a message in an incredibly fast relay-race fashion. It's thanks to them that when you stub your toe, your brain quickly gets the signal ("ouch!"). It's also why you start hopping on one foot and saying words you wish your kids didn't hear (the brain promoting uninhibited behavior, again by neurotransmitters). Microbes seem to have figured out this concept, and whether they make or modify our neurotransmitters, they can rapidly join in the conversations. Although there are a few examples of microbes specifically producing neurotransmitters, this has been observed only when they were grown in a test tube, and there are no known examples of them actually doing this in our guts (yet). However, there are many examples of the microbiota affecting our body's neurotransmitter levels, which can in turn affect how our brain works. This is beginning to provide molecular clues as to how the microbiota may affect people's behavior, depression, and stress.

This area could be an entire book on its own (actually, there is at least one), and many neurotransmitters with complicated names and abbreviations such as BDNF, dopamine, GABA, G-CSF, serotonin, acetylcholine, and norepinephrine are all affected by the microbiota. All play major roles in our everyday brain and nervous system functions, and alterations in these neurotransmitters can lead to neurological problems and mental illnesses. Instead of going through all these mechanisms in detail, we'll use just one as an example to illustrate how things might work.

Let's pretend you and your spouse hire a babysitter and go out to a nice restaurant for a quiet dinner (finally!). Picture that satisfied feeling of being able to finish your meal without interruptions, while it's still warm! The reason you feel that way is due to a neurotransmitter called serotonin (plus the fact that you're not witnessing your two kids fighting with each other—that's the babysitter's problem). Nearly all of your body's serotonin is made by cells in the gut, where it regulates intestinal movements (needed for proper gut functions). In the brain, serotonin regulates mood, appetite, and even sleep. Serotonin is made from tryptophan, a building block of proteins. So if tryptophan levels are affected, so are serotonin levels. How do the microbes dial in? They affect tryptophan levels, as well as the chemicals that control serotonin synthesis. In germ-free (GF) animals, serotonin levels are decreased in both the blood and the colon, and this can be corrected by adding back certain microbes to the animals. These microbes make small molecules (metabolites) that affect serotonin production, and if you add these metabolites back to GF animals, even this can restore serotonin levels.

Experiments with GF mice have been quite helpful in demonstrating the large impact the microbiota has on brain function. For example, GF mice are more exploratory, less anxious, and significantly less fearful than mice born and raised with regular microbe-rich conditions (not good from a mouse's point of view, but great news for a cat!). It's believed that certain microbes regulate the secretion of an important neurotransmitter molecule, known as GABA (gamma-aminobutyric acid), demonstrating yet another critical human function delegated to microbes.

GABA is necessary to inhibit feelings of fear and anxiety, two essential emotions that all animals need to control as a primal survival strategy. Amazingly, colonizing GF mice with bacteria, or even doing fecal transfers from a normal mouse to a bold GF one restores their behavior back to normal. However, this is time-limited, as only GF animals recolonized early in their lives can restore normal behavior, emphasizing once again the importance of microbes early in life.

Other neurological aspects that are altered in GF mice are cognitive functions, learning, memory, and decision-making, which are all pretty important things when thinking about childhood development.

The above-mentioned experiments were done in the absence of microbes, which, as we have mentioned before, is an unrealistic scenario. However, there have been other behavior-changing experiments done in the presence of *different* microbes. For example, feeding lab flies a specific bacterium (a probiotic) makes them prefer to mate with other flies colonized with this particular microbe, while they avoid potential mates that lack this bug. It's thought that the bacterium provides building blocks to molecules called pheromones that are scent attractants.

Similarly, hyenas hang out in tribes or social groups, much like teens hang out in cliques. These animals have scent glands (which stink, unless you're a hyena), and specific tribes have their own set of microbiota living in these scent glands that affect which pheromones are made. While this has implications in hyena tribal structure, it has even bigger implications in that the microbes are affecting mate selection, which could actually drive evolution.

Microbes and Moods

At present little is known about how microbes directly affect human behavior, but given the tantalizing animal data, studies are now under way. With that said, other clues emphasize that the microbiota is needed for a healthy mental state in humans. Many psychiatric illnesses and mood disorders are associated with gastrointestinal disorders. For example, depression is common in people with irritable bowel syndrome and they are sometimes treated with antidepressants.

In any given supermarket, there always seems to be a two-year-old having a temper tantrum (perhaps even your own child). Although not a psychiatric condition (many parents might disagree), the unruly and defiant behavior of most two-year-olds is a developmental stage to be reckoned with. For most parents, the so-called terrible twos is something one doesn't easily forget. Going from a sweet, smiling child to a wailing, flailing irrational toddler is a remarkable change in behavior. Luckily, it's a time-limited event, and kids outgrow it.

Could the microbiota play a role in this? There's very little data at this point, but one recent study of the microbiota in children between 18–27 months showed that the presence of certain bacteria is associated with a reduced ability to control emotions, especially in boys. We also know this is a time in a child's life when remarkable changes are occurring in their microbiota as they shift from being an infant to a toddler. Thus, it's entirely possible that bacteria may be involved in this dramatic shift in temperament. This doesn't mean that the microbiota is causing those changes; it may very well be that the microbiota is responding to changes in stress-related hormones. Either way, the next time your toddler is rolling around in the supermarket aisle desperately demanding a toy, you can practice patience by thinking perhaps you can blame it on your toddler's microbes and not on your parenting skills.

If microbes are so good at interacting with our brain, could they actually affect the way it develops and functions? Recent studies suggest that early-life changes in the microbiota may impair memory. Experiments that tried to mimic the Western diet of developed countries showed that if you put mice on a high-sugar diet early in their development, it affects both long- and short-term memory later in life. It also causes changes in the composition of the microbiota, which led the authors of the experiment to state: "These results suggest that changes in the microbiome may contribute to cognitive changes associated with eating a Western diet." Scary stuff, when considering so many children in our society follow a high-sugar diet.

In other experiments, in which the microbiota was modified by an infection instead of diet, mice infected with a harmful microbe and then subjected to stress showed memory dysfunction, which could then be prevented by probiotic treatment. Additional experiments with GF mice showed that their memory was impaired compared to normal mice, whether they were stressed or not, indicating that our resident bacteria somehow impact the capacity to memorize and remember. Still, without human data it's far too early to definitively say whether the microbiota directly impacts human memory or other aspects of intelligence, but there are some tantalizing clues that could have profound societal implications if they prove true.

Stress, Depression, and Anxiety

We live in a stressful world, and this has all sorts of implications for our microbiota and brain. It starts in the womb. For example, maternal stress and infections during pregnancy are linked to neurological disorders such as schizophrenia and autism spectrum disorder in children. It's thought that the ability to cope with stress is set up early in one's life. This system is immature at birth, and goes on to develop at a time when the microbiota is changing rapidly. Stress during this formative time is also linked to later brain disorders.

Moreover, early-life stress changes the microbiota. We know that the separation of rats from their mothers early in life leads to long-term changes in their ability to deal with stress, as well as long-term changes to their microbiota. Adult GF mice show exaggerated stress responses that can be fixed if microbes are encountered early in life, but this cannot be reversed if microbes are added later. However, when mice were fed a particular probiotic, these animals were less stressed. In fact, the effect of the probiotic was as strong as giving them antianxiety medication. They also made more cortisol, a hormone that helps cope with stress. If the vagus nerve connection was severed, the effect was lost, again implying a strong gut-brain connection. Similarly, GF mice have decreased cortisol, which helps explain why they have increased stress reactivity. This is obviously a new and experimental field, with nearly all the results coming from animal studies. However, it does have significant implications for raising children in a stressful world. It's interesting to contemplate a world in which an infant's microbiota could be modified to improve their ability to deal with stress and anxiety later in life.

Depression and anxiety may also have links with the microbiota. One school of thought is that depression is an inflammatory disorder (which would affect the microbiota). Damage to the gut, which causes inflammation, has been linked to depression (as well as schizophrenia and autism spectrum disorder). Incredibly, anxiety can be transmitted in mice simply by doing a fecal transfer between animals. In a small human study of fifty-five people (thirty-seven had depressive disorders and eighteen were controls), it was found that there were several correlations between emotions and changes in the microbiota. Larger studies addressing this are warranted, and fortunately these are now under way.

Some probiotics have shown beneficial effects by decreasing depression-like behavior to the same extent as antidepressant medication, but again, this has been studied only in animal models thus far.

Autism Spectrum Disorders

Autism, a neurological disorder affecting more and more children every year, has deservedly received a lot of public attention. Autism spectrum disorders (ASD) are a collection of neurodevelopmental disorders characterized by social and communication difficulties, repetitive behaviors, and sometimes cognitive delays. They include autism, Asperger's syndrome, and other related disorders. Unfortunately, the rates of these disorders are increasing; the World Health Organization estimates that 1 in 160 children have ASD and this number continues to increase. Because these rates have risen so rapidly, it's unlikely that it's caused by genetic changes alone. A common reason cited for the rise in autism is the increased diagnosis of ASD—the argument that we simply weren't identifying the disorder before—but again this doesn't fully account for such a steep rise (and where are all the autistic sixty-year-olds now that weren't diagnosed long ago?). In any case, the debate over whether it's an increase in diagnoses or in the actual number of cases is irrelevant when there's an epidemic of ASD affecting such a large proportion of children.

Though theories abound, there's no known cause of ASD. However, links to the microbiota have recently gained much attention. It's been well established that maternal stress and infections during pregnancy are linked to neurological disorders, including ASD and, as we previously discussed, this is accompanied by microbiota changes. Children with ASD often have serious gastrointestinal problems, such as diarrhea, constipation, bloating, abdominal pain, and increased intestinal permeability, all of which could affect the microbiota–gut–brain axis. These symptoms are especially present in children diagnosed with regressive-onset ASD (children who were developing normally, but started showing symptoms as toddlers, around 15–30 months of age) with abnormal behavior and loss of previously attained skills. Furthermore, several reports suggest that the microbiota of ASD children is different from the normal control groups. One hypothesis being put forward is that ASD may be either caused or enhanced by a general imbalance of the microbiota.

Based on this hypothesis, there was a small study in which the antibiotic vancomycin was given to ten children with regressive-onset autism who had also manifested gastrointestinal symptoms. Eight of these children showed clear improvements after treatment, but these were short-lived. In a similar study in France, 80 percent of regressive-onset autistic children reported dramatic improvement in their symptoms but, as with the previous study, it didn't last long. Could it be that once the antibiotic treatment stopped, the microbiota reverted back to its previous altered state? The following compelling real-life story suggests that this just might be the case.

Leah was a normal little girl who at five years of age developed an infection, as many five-year-olds do. A week later, her parents started noticing very odd and serious symptoms and took her back to her pediatrician, who became worried after seeing her autistic-like behaviors. Leah had begun walking on her toes, flapping her hands constantly, screaming, avoiding eye contact, showing minimal verbal skills, etc. As parents ourselves, we can only imagine the severe anguish her parents must have felt after seeing their healthy daughter's personality transform this way in just a few days.

Her father, a university professor, sat in front of the computer for days and gathered all the information he could, then went back to Leah's doctor. He managed to convince him to prescribe Leah with a short dose of the antibiotic vancomycin to change her microbiota. Antibiotic treatment produced a night-and-day effect in this little girl. While on vancomycin her symptoms almost disappeared, yet off vancomycin the autistic symptoms returned! At this point they were convinced they could treat Leah's symptoms by modifying her gut bacteria, but they couldn't keep Leah on antibiotics all her life, so they needed to find a better solution.

Leah's father contacted the original scientist who had published about this and he suggested they find a doctor who would be willing to do a fecal microbiota transplant, or FMT. As the name suggests, this procedure involves the transplantation of fecal matter from a healthy individual into a recipient. Unfortunately, it was exceedingly difficult to find someone in the US who would do this procedure at that time, so the family traveled to Canada, where they had managed to convince a gastroenterologist to do the procedure (try telling that story to the border guards when asked if you have anything to declare). Leah responded incredibly well to the transplant, not showing any symptoms for months. She did have a mild relapse after another infection, but an additional FMT made those disappear again.

The scientist and doctor involved in Leah's case analyzed her fecal samples before and after the FMT and found that her microbiota had definitely changed. Interestingly, her sample before treatment showed an unusually high number of a bacterium called *Clostridium bolteae*, a microbe that had been previously associated with autism. The story of the discovery of this bacterium is very unusual, but hopeful.

Let's also look at the case of Andrew. He's now twenty-three, but when he was about eighteen months old, he started showing signs of autism and severe gastrointestinal issues. He had received numerous bouts of antibiotics to treat fluid in his ear over the span of three months. During the sixth course of antibiotics his mother, Ellen, started noticing significant behavioral changes. Andrew became withdrawn, not wanting to be held or touched, even by her. Ellen soon suspected there might be a connection between the antibiotics and Andrew's altered behavior. Just like so many parents whose children have been diagnosed with regressive-onset autism, Ellen bounced from doctor to doctor, and was met with a lot of disbelief and skepticism. So, just like many parents in this same situation, she devoted herself to investigating if there was a link between antibiotic treatments and autism. However, unlike most parents, Ellen became so dedicated to her research that in 1999, and without any formal scientific training, she published a paper proposing that autism could be a disease of microbial origins. When Andrew was five, she decided to take all her research findings and her own publication to a pediatric gastroenterologist—the thirty-seventh doctor she had seen since Andrew was diagnosed. It was this doctor who finally agreed to prescribe Andrew an eight-week treatment course of vancomycin.

Ellen and her doctor started documenting the effects this antibiotic had on Andrew through videos and additional doctor visits.

The changes immediately after starting the treatment were overwhelmingly positive. Within a couple of weeks Andrew was using speech again, he could wear clothes, he was hugging his mom, he was even able to potty train in the span of a week. As Ellen eloquently puts it, "Andrew was aware of his surroundings again—he was starting to come back." Andrew's doctor could not believe that the child who had previously been so irritated and out-of-control in former visits was the same child now patiently sitting down waiting for the doctor to see him.

After seeing how successful Andrew's treatment was, a clinical trial was immediately started. However, as the weeks passed, Ellen's enormous relief and happiness turned into heartbreak, since at the end of the antibiotic treatment Andrew's behavior started spiraling downward again. Nevertheless, they had discovered that his autistic symptoms could be modified with an antibiotic, something completely unrecognized before.

Her efforts of course didn't stop there; she became involved in several clinical studies led by Dr. Sydney Finegold, a medical researcher from the University of California, Los Angeles. Together, they discovered that children with regressive autism often had an unusually high abundance of a previously unknown bacterium, which they named *Clostridium bolteae*, in honor of Ellen's heroic efforts for this cause (Ellen's last name is Bolte). These bacteria are extremely resilient to antibiotic treatment, and once the treatment stops they always come back.

Since then, Ellen has tried to modify Andrew's microbiota through other methods such as diet and strong doses of probiotics. Recently, they subjected Andrew to an FMT, which unfortunately did not improve his autism. They believe it didn't work because it was done when he was twenty years old, a very long a time after his symptoms started. They speculate that over time the processes affecting the nervous system that responded to a microbiota treatment might have become unchangeable. Still, the severity of Andrew's autism has decreased dramatically from the time they started treating his intestinal microbiota. He now has a much less severe form of autism, something unusual in ASD.

These stories are truly compelling, but it is important to emphasize that it has not been proven yet that this or any other bacteria *cause* autism. It could be that the microbiota of children diagnosed with ASD change due to the elevated anxiety and dietary changes that these children undergo. More importantly, not every case of autism improves after antibiotic treatment, likely because ASD includes a large group of disorders with similar symptoms but different causes. Parents of children diagnosed with regressive autism should definitely consider bringing up this information with their health practitioner, yet at the same time refrain from becoming overly optimistic that this type of treatment will work for their particular child.

A lot more human research is necessary to determine the true role of certain gut bacteria in ASD development. However, there are very persuasive arguments about microbiota involvement in ASD coming from, yet again, germ-free (GF) animal studies. In a well-known study from Sarkis Mazmanian's lab at the California Institute of Technology, researchers used a mouse model of ASD to probe the role of the microbiota. As in humans, pregnant mice exposed to stress produce offspring with some of the behavioral and physiological changes seen in ASD, including repetitive movements like obsessively burying marbles, and showing less communication and fewer interactions with other mice. They also found that the microbiota in mice with ASD-like symptoms was altered and they had increased gut permeability, just as in children with ASD. Digging deeper, the scientists identified a small molecule that was more prevalent in the ASD mice and that is similar to one found in human ASD. This molecule, which is produced by microbes in the intestine, triggered ASD symptoms when given alone to mice, suggesting for the first time that a bacterial compound can cause or at least trigger ASD in mice. They went on to suggest that because of the increase in intestinal permeability in ASD mice, this molecule could also get to the brain more easily. In support of this idea, when they gave a probiotic known to decrease intestinal permeability, it restored the microbiota, decreased the production of this metabolite, and remarkably, also decreased the ASD symptoms in these animals.

So where do we stand regarding microbiota and ASD in children today? There are an awful lot of smoking guns hinting that the microbiota might be involved, as we have seen in the above cases. However, nothing has been proven in humans. Neurologists are now aware of these many correlations, and several studies are being undertaken to test this hypothesis in humans, but it's not yet common medical practice to consider the gut microbiota when treating ASD. It seems strange indeed to consider that something as simple as a probiotic or a fecal transfer could influence such complex neurological disorders such as ASD, but given the animal studies and many other human correlations, it may not be such a far-fetched proposition if we get the probiotic strain(s) right.

Attention Deficit Hyperactivity Disorder

Another childhood behavioral disorder that's becoming very common is Attention Deficit Hyperactivity Disorder (ADHD), with up to 12 percent of boys and 5 percent of girls being diagnosed with it in the US. The average age of diagnosis is seven years, although these children are often hyperactive in the womb. This disorder is characterized by lack of attention, impulsiveness, and hyperactivity. Like ASD, this disorder has a big spectrum and wide degrees of severity. There are children who show mild symptoms and children with very severe manifestations that prevent them from attending school or carrying on a normal life. Such kids may tend towards risky behavior as teens, including drunk driving, substance abuse, unprotected sex, etc.

Many factors are thought to contribute to ADHD, including a genetic component (it can run in families) and factors encountered during pregnancy, such as low birth weight, prematurity, and prenatal exposure to alcohol and smoking. Is there a microbiota link? We don't know for certain, but again there are several hints. Children with food allergies, eczema, or asthma (all associated with microbiota) have increased rates of ADHD. We also know that diet changes can sometimes reduce the symptoms of ADHD, which would also affect the microbiota.

In a remarkable (but very small) study, forty children were given a probiotic for the first six months of their life, while a placebo group of thirty-five children were not given the probiotic. Thirteen years later (it was a long experiment!), they found that 17 percent of the control children had ADHD and/or Asperger's syndrome (three had ADHD, one had Asperger's, and two children had both), while remarkably not one child in the probiotic group had either of these disorders (o percent)! They did look at the microbiota, but could not define a single microbe that correlated with this, although thirteen years ago the tools available for microbiota analysis were not very good. We want to stress again these are very small numbers and only one trial, but they are extraordinary if indeed they hold true.

The Road to a Better Brain

Can we make our brains even healthier, or fix neurological problems by changing the microbiota? Maybe. There are currently three main methods that seem to show promise in improving brain health. The first is diet (which of course also alters the microbiota). We've all heard about the beneficial effects of omega-3 fatty acids, and eating lots of healthy foods such as fruits and increased fiber (see Food for Thought on page 223). The second way is exercise. Studies have shown that exercise, even in small amounts, has anti-inflammatory effects. It also leads to changes in the microbiota that appear to be beneficial. Perhaps there's a reason kids roar around so much. The final way, which is being studied extensively, is to more directly modify the microbiota using antibiotics, prebiotics, and fecal transfers. However, most of the focus is on probiotics. These living microbes seem to have beneficial effects, and if they're used for affecting the brain, they now have their own new name called "psychobiotics."

Probiotics are a huge market worldwide, estimated at over $20 billion in sales. Despite their large consumption, there have been only a few human studies done on their effects on the brain. However, there is some very compelling data coming from animal experiments that we should look at before we briefly discuss what's known in humans. For example, if mice are given a probiotic for twenty-eight days, it decreases their anxiety-like behavior. Similar experiments were also performed on rats, and one study found that two probiotics given together had the equivalent effect to antianxiety medication. In other rat and mouse models, scientists found that probiotics decreased depression to the same extent as antidepressants. Studies have also shown probiotics to improve learning and memory in mice.

Unfortunately, in humans there just aren't sufficient studies yet to conclusively recommend a particular probiotic for certain conditions or brain health. There is preliminary research that suggests probiotics may help cognition and stress-related conditions such as anxiety, autism, depression, and schizophrenia, and even multiple sclerosis. Great! We should all take probiotics, including our kids, right? Well you certainly can, given that probiotics are safe and will not cause an adverse reaction in children. The problem is we don't really know which ones do what in people, and whether they really work for neurological disorders. The bottom line is that there are some very promising results done in very small trials, but what we need are larger well-designed clinical studies to figure out whether they really do work and which probiotic(s) should be used and for what.

Dos and Don'ts

- ◆ **Do—** get your kids to eat healthily and exercise regularly, since studies show how this benefits the gut microbiota, as well as improves brain development (they're probably linked). Remember that a healthier gut means a healthier brain, as the microbiota have much more to do with our brain than we previously thought.

◆ **Don't—** worry, be happy! Same for your kids. Easier said than done, but we do know that stress has a detrimental effect on the gut microbiota, which in turn can affect the brain.

◆ **Do—** follow the press and literature about probiotics, and look for large-scale controlled clinical studies that suggest they work. The field is currently very active, and could change quickly in the next few years. It often takes a long time to move from a successful trial into regular medical practice. This is especially true for more "unconventional" treatments such a probiotics, given their history and lack of regulation. It's also going to take a while to convince a neurologist or psychologist to treat a neurological disease with intestinal microbes.

◆ **Do—** consult with your child's pediatrician or psychiatrist, if she suffers from ASD or ADHD, about their opinion on the impact of the microbiota, but please don't expect micro-biota alterations to be a certain cure. Consider consulting a different health practitioner, such as a gastroenterologist or a naturopath, who has been involved in treating ASD or ADHD patients. They may agree to try certain minimal risk options, such as antibiotics, probiotic treatment, or a fecal microbiota transplant (for ASD).

FOOD FOR THOUGHT

C an we improve brain function in children by feeding them certain foods? We know that malnourished children have decreased cognitive function, presumably because of nutrient deprivation, hampering full brain development.

Diet plays a major role in maintaining brain health in older people, as well. Eating plenty of plant fiber and other antioxidants decreases the risk of dementia and other neurological diseases such as Alzheimer's and Parkinson's later in life (if this interests you, you might read *Brainmaker* by David Perlmutter).

We also know that the brains of germ-free mice don't develop normally. So obviously good nutrition and microbes are key for proper brain function. But can you actually improve the brain of your child through diet? There are no definitive answers yet, but given all that we're learning about microbes, diet, and brain function, a fascinating (but very controversial) study would be comparing microbiota composition, diet, and IQ scores. Until someone does the experiment and finds out if there are "smart bugs," make sure your child eats as healthfully as possible, including lots of fruits and vegetables—just tell them it's "brain candy!"

15: Vaccines Work!

The Not-So-Magical Kingdom

Five-year-old Ethan was so excited he barely slept that night, anxiously tossing and turning, and waiting for the first rays of morning sunshine—he was going to Disneyland! He could finally see the big castle where Mickey Mouse lived, and check out the Pirates of the Caribbean (he had been obsessed with this movie lately and wore his pirate hat everywhere). At last, his mom and dad pushed his bedroom door open and he sat up like a spring. However, his giant smile was met by his parents' concerned faces. His father sat on his bed and gently said, "Ethan, we're very sorry, but we can't go to Disneyland right now."

On a tantrum scale of one to ten, Ethan threw an eleven. Finally, when Ethan had calmed down, he was able to listen to the reasons his parents were trying to give him: "There's a very bad disease called measles that is infecting children at Disneyland right now. You're protected from it, but your sister Olivia isn't old enough yet and she could get very, very sick." Ethan's parents were right; Olivia was only nine months, not old enough to be fully vaccinated and protected against measles. "We have to stay home," they said. To which Ethan promptly answered: "It's not faaaaaaiiiiiiiiir!"

Sure enough, during the spring of 2015 a measles outbreak that started at Disneyland infected 131 children. Disneyland was, for the first time ever, recommending families not to come unless every visitor was fully vaccinated. It was simply too risky. That winter, the Happiest Place on Earth was not quite so.

Measles is an extremely contagious childhood disease. One infected child can infect another eighteen kids, and approximately 90 percent of unvaccinated people will contract the disease if exposed to the virus. Measles is miserable! Fever, cough, runny red-rimmed eyes, and an extensive rash are the main symptoms. In some cases, measles cause lung infections (pneumonia), brain infections (encephalitis), and even death (1 in 2000 cases). The CDC calls it "the deadliest of all childhood rash/fever illnesses."

Fortunately, since 1970 there has been a childhood vaccine that has nearly eradicated this nasty disease, which affected 900,000 people per year prior to vaccination in the US. The vaccine is one part of the MMR vaccine (measles, mumps, and rubella) that's given to children at 12–15 months of age, and then a booster shot is given again at 4–6 years of age. Those of us born prior to 1970 will never forget how miserable this childhood disease is, as nearly all children suffered through it. The vaccine is produced by crippling a live virus to the point that it can't infect very well, yet it can still "tickle" the immune system and cause it to "remember" the virus. The immune system is very good at remembering these things, and if a vaccinated child is exposed to measles, he is protected 98 percent of the time.

Vaccines, along with antibiotics and sanitation, have been fantastic tools to combat infectious diseases. Vaccination has been remarkably successful, and has rid the world completely of smallpox, and very nearly of polio. Ironically, their success has also been their downfall, as living without these diseases have led some people to believe that vaccines aren't that necessary.

A year before the Disneyland outbreak, in the spring of 2014, an area just east of Vancouver, Canada, called the Fraser Valley, started to see cases of measles. Within four weeks there were over four hundred cases of measles, which is more than the province of British Columbia had seen in the last fifteen years combined and the worst outbreak in thirty years. When investigators started looking closely at the cause, they found that the cases were clustered at a Christian school (which had to be closed for a while) that's populated by the Reformed Congregation of North America. This group does not believe in vaccines, saying they're not safe and citing other religious reasons. Consequently, this area has a high rate of unvaccinated children. Fortunately, most of the surrounding area (and most of Canada) has over 90 percent vaccine compliance rates, which prevented further spread of the disease.

Despite this outbreak, according to one survey, 80 percent of anti-vaccine parents remain "not at all likely" to vaccinate against these nasty childhood diseases, citing health and religious concerns. You could say, "Fine, that's their choice." The problem is, it could also affect your kids, because to eliminate a disease from a population and break the infection cycle, the large majority has to be vaccinated. Children younger than twelve months, like Olivia, are not vaccinated and are at risk. In addition, the vaccine doesn't "take" in some kids, leaving a portion of people unprotected from the disease.

A Parent's Nightmare—What Do I Do?

A situation like this hit too close to home a few years ago, with a different preventable disease. Claire's son was only six weeks old when her husband took him to a doctor's appointment. It was her husband who needed to see the doctor, but he was in charge of their son that morning, so he came along for the ride. Strapped in his car seat, he slept through the whole thing like only a six-week-old can. There was nothing out of the ordinary until three days later, when the doctor's receptionist phoned them at home. She said that they had to bring their son in immediately to see the doctor because he had been exposed to whooping cough. It turned out that a different doctor from the same clinic had seen two school-age siblings with whooping cough an hour before Claire's husband's appointment. Claire assured the receptionist that her son was doing great and that he hadn't even been examined that day. Still, "Your son was here, I remember seeing him, this is why I'm calling you," she insisted. She said that because he was so young, he must be treated with antibiotics to prevent a possible infection.

Claire got dressed, asked her neighbor to look after her two-year-old daughter, and took her son to see the doctor. While in the waiting room, Claire read recent studies on whooping cough. She was familiar with the disease, but wanted to be more informed before she saw the doctor. Whooping cough, also known as pertussis, is transmitted by the bacterium *Bordetella pertussis*, a very contagious pathogen that causes severe coughing and difficulty breathing. In older children and adults it can last a long time (often called the hundred-day cough). In adults it's usually treated at home, often requiring antibiotics. However, infants often require hospitalization, and in this age group complications include pneumonia, seizures, brain damage, and even death.

Once Claire saw the doctor, he explained that the risk of contagion was extremely low, as her son never left the car seat, and he was never in the same room where the infected children were examined. His potential exposure came from being in the same large waiting room where these children had been an hour before. Although the doctor assured her that it was incredibly unlikely that her son would be infected, provincial health policy required him to recommend and prescribe her son a preventative dose of antibiotics. Claire wasn't very sure she agreed with what she was being told. So, right before the end of the consult, she asked him if he would give the antibiotics to one of his own children, to which he answered that he probably wouldn't. Once she got back to the car she opened the doctor's prescription note and read: *azithromycin for ten days*.

Now, Claire is not the type of patient to refuse medical treatment, but in this case she needed a second opinion. After all, her child seemed perfectly healthy and the risk of infection was minimal. Not only did she consider it unnecessary, she was obviously very aware of what ten days of azithromycin could do to her son's microbiota. She was already immersed in the microbiota field, reading study after study showing how antibiotics early in life increase the risk for all sorts of diseases. Plus, Claire's son was already genetically predisposed to asthma (with very severe cases of the illness in the immediate family), so the last thing he needed was a big shift in his microbiota at such a young age.

But after consulting with no less than six other physicians and researchers (mostly friends and friends of the family), the recommendations were completely mixed regarding whether or not Claire should give her son the antibiotics. So Claire came up with plan B:

for the next five days she was going to monitor her son's temperature every four hours around the clock. If at any point there was a small increase in temperature, she would start him on antibiotics. She followed the plan meticulously (just like doing an experiment), which made for five very long days and nights that she will likely always remember.

Fortunately, her son had not been exposed and he never developed whooping cough. Yet to this day and after imagining all possible scenarios, she still doesn't know if she made the right decision or not. As a mother of young children Claire has become familiar with the idea that parenting comes hand-in-hand with making decisions that you might regret later. However, in this particular case she knew that she shouldn't have been put in this position.

This happened in December 2012, a winter that saw the worst whooping cough epidemic in the Pacific Northwest since 1942. It was that day, waiting for Claire's doctor to see them, that she really understood how dangerously successful the anti-vaccination movement was. Those two kids in the waiting room had not been vaccinated, and as a result were a real threat to her son. More and more people are becoming convinced that growing up without vaccines is a safer option than getting vaccinated. How can this be? How can a parent believe that the risk of an extremely rare vaccine reaction is bigger than the risk of contracting a very dangerous infection?

Unlike popular belief from pro-vaccination activists, we don't think "anti-vaxxers" are acting out of a lack of information or mere stupidity. On the contrary, a quick online search on why some people choose not to vaccinate their children yields not hundreds, but thousands of articles with scary stories of vaccine reactions and reasons to avoid vaccines. In fact, some of the information we saw was somewhat compelling, and if we didn't have a strong scientific background it wouldn't be that hard to succumb to these theories. After all, Western lifestyle diseases are increasing dramatically and they are affecting more and younger children, and something must be causing this. As one parent said, "How can we believe vaccines don't cause autism, if science hasn't told us what does?" A very valid question from someone who is dealing with the overwhelming reality of having an autistic child, and who still has no answers to why this terrible disease is on the rise (see chapter 14 for a discussion on autism spectrum disorders).

Unfortunately, not knowing an answer does not make a wrong statement right, and the science behind anti-vaccination theories is not solid. There just isn't a single validated study that, as scientists, we can believe shows that vaccines cause diseases. Yes, vaccines are associated with certain reactions, and in very exceptional cases they can be severe, but in the end, it comes down to assessing risk. According to the World Health Organization, the risk associated with severe neurological reactions to the DPT vaccine (against diphtheria, polio, and pertussis) is extremely low—it occurs once in every 5 million cases. Compare that to the risk of your child developing whooping cough in the state of Washington, where there are 62 cases per 100,000 residents, and the number continues to rise every year. That is one thousand times more likely. Moreover, add the risk of not only having your child suffer a severe infection, but also passing this infection to other kids, who are either too young or too sick to be vaccinated, like a six-week-old visiting a health clinic.

We're currently parenting during the digital age, where any information we want (and don't want) is at the tip of our fingers. It's not easy to make decisions with this amount of information around, but please, don't fall for the quick-access blogs and "health-oriented" articles promoting an all-natural approach to protecting your children from vaccines. Not unless you decide to move to the middle of the woods, away from society. Infectious diseases are a reality of living in large groups of people; they have been around as long as we have. The only reason our children don't suffer from them now is because of vaccines, and without vaccination there's no other alternative than having these diseases come back. So, despite extremely rare cases in which vaccines cause severe reactions, vaccines work and are one of the safest medications in the world.

Vaccines and Microbiota — Is there a Connection?

Vaccines work, yes, but they are not perfect. Why don't vaccines protect 100 percent of the people that take them? Most vaccines work about 85–98 percent of the time, which leaves a significant number of people semi- or unprotected. A good example is the seasonal flu vaccine. We all know people who get the vaccine, but then still get the flu. We also know that vaccines that work really well in developed countries often work poorly in less developed places where these diseases have a major toll on children. This is true for polio, rotavirus, and cholera vaccines. Remember, these children have a very different microbiota—could this be influencing their vaccine responses? The microbiota has an important role to play in the way our immune system functions.

As an example, when animals are given antibiotics their antibody response changes. Antibodies are an essential aspect of a vaccine response and antibiotics are a great method to shift the microbiota. Furthermore, animals that are raised germ-free have poor responses to vaccines, and feeding probiotics or prebiotics to mice affects their subsequent responses to vaccines. All these results suggest that the microbiota may affect vaccine responses.

This makes sense, as the microbiota is critical for your child's immune system to develop normally. In fact, specific microbes have been identified that shift the immune response in different ways. Throughout this book we've seen that our microbiota plays an important role in many diseases because of their ability to influence our immune system. From this perspective, the microbiota can also influence a vaccine response just as it does with other aspects of our immune function.

Scientifically understanding the role of the microbiota in vaccine responses is in its very early days. There are only a few experiments thus far, although they are increasing in number. We do know that in people given a typhoid fever vaccine, the vaccine did not affect the microbiota. Although we expected it, this is good news. The microbe that causes typhoid fever (a type of *Salmonella*) is not part of the normal microbiota, and is usually only present when it causes disease, so the vaccine targets that microbe only. However, this does raise a bigger question—what happens when we take a microbial species out of the normal population? Does this change the overall microbiota structure? Could something worse crawl in?

The experiment has been done once, with the elimination of the smallpox virus from the world, and luckily this scenario has not happened—no new nasty virus has appeared in its place. With the recent introduction of pneumococcal (pneumonia) vaccines, which target fairly common microbes of the respiratory tract that cause ear infections, we will be watching closely to see what, if anything, replaces them. In experiments done on macaque monkeys, it was found that monkeys with the most diverse microbiota responded the best to certain diarrheal vaccines. We presume a similar concept holds true for children, again arguing for letting our kids eat dirt, and whatever else they can safely put in their mouths. It seems that having a diverse microbiota is beneficial all-around and we should provide our children with opportunities to diversify their microbiota.

What about the less fortunate parts of the world? There are major groups, such as the Gates Foundation and Gavi, The Vaccine Alliance, that are trying to get as many children as possible vaccinated against common childhood diseases. However, as mentioned above, children in developing countries do not respond as well to vaccines as those in developed countries (where the vaccines are often developed and tested).

With the aim to study how to improve this, our laboratory has developed a mouse model that mimics some of the conditions observed in developing countries. By shifting their diets to have more carbohydrates and less protein and fats (similar to certain childhood diets in developing countries), and then feeding specific intestinal microbes to these mice (kids in developing countries live in less hygienic conditions), these mice have remarkably similar symptoms to malnourished children—they have stunted growth, intestinal inflammation, chronic diarrhea, poor development, and all the things one normally sees in these children. Intriguingly, these animals have very different immune responses, too. This puts us in a position to experimentally examine how the microbiota affects the immune response, hopefully enabling us to work on making even better vaccines and delivery systems for impoverished children of this world.

As mentioned before, besides not being 100 percent effective, vaccines, like any medicine, are not 100 percent safe. Sometimes there can be side effects, although rarely are they serious. Most of the side effects are minor—pain (hey, needles hurt), swelling, and redness. As anyone who has taken a first aid course knows, these are the cardinal signs of inflammation. Ironically, this is probably good, as it means you're tweaking the immune system, and it needs to activate and remember it (much like one certainly remembers better if you are kicked in the shins, rather than gently tapped on the shoulder). Fever and irritability can accompany vaccination, although these can be dealt with fairly easily. It's the serious complications such as seizures and even a risk of dying that one worries about. Using the MMR vaccine as an example, there has been just one death in more than fourteen years, although it wasn't directly attributed to the vaccine. Serious events for most vaccines average about one in a million. However, before the MMR vaccine was introduced, one in a thousand kids died of measles, which means there was a thousand-fold higher chance of dying before the measles vaccine was developed. The problem is that we don't see children dying from these diseases because of vaccines, so even one in a million odds seem high when dealing with your own child.

How could the microbiota play into adverse vaccine events? Again, we know the microbiota impact on our immune system development, which is tweaked by vaccines. There's currently no information either way about how microbiota might affect such outcomes, but given everything we're learning about how much these microbes do in our body, there's certainly a possibility that they might be part of the equation. Time will tell. Until then, it's best to act wisely and vaccinate our children and try our best to keep their microbiota healthy.

Dos and Don'ts

- **Do—** get your child vaccinated according to the standard guidelines (these vary by state, province, and country—the standard vaccine schedules for different areas are all available online). If you remain skeptical about vaccines, have a serious discussion with your pediatrician about the risks associated with childhood vaccines.

- **Don't—** believe everything you read on the web about vaccines causing autism and other diseases. Some of this is based on fraudulent science that has been retracted, and other hype that is not based on science at all. There is a very small risk of adverse reactions, but these are minimal compared to the problems caused by the actual disease. Talk to your doctor, or your grandparents that lived before vaccines—they were terrified of polio and other diseases we no longer even hear about.

- **Do—** follow our advice throughout this book on how to maintain your children's microbiota in good health. A diverse microbiota promotes healthy immune responses, and increases resistance to infectious diseases.

A MYTH THAT HAS LASTED FAR TOO LONG

The MMR vaccine is a highly effective vaccine that is routinely given to all children as part of the normal vaccine schedule to prevent measles, mumps, and rubella. In 1998, Andrew Wakefield and colleagues published a paper in the prestigious medical journal the *Lancet*, suggesting that the MMR vaccine causes autism spectrum disorder (autism), based on a very small study of twelve children. The media picked up the story, and very rapidly the rates of vaccinations in the UK and Ireland dropped, resulting in a jump in measles and mumps cases, along with the deaths and long-term damage associated with these diseases.

Unfortunately, the scientific study was a fraud. Wakefield did not declare the numerous conflicts of interest regarding his sources of funding, plus he manipulated evidence and breached several ethical codes. He was stripped of his medical license in the UK. As study after study came out showing no correlation between MMR vaccines and autism, the *Lancet* partially retracted the paper in 2004, and fully retracted it in 2010.

At least a dozen studies since have shown no correlation between autism and the MMR vaccine. The latest, published in 2015, looked at 95,000 US children, including younger siblings of autistic kids (who have a higher chance of getting autism), and again concluded there was no correlation. One medical journal article called the original study "perhaps the most damaging medical hoax in the last 100 years."

The question is: Why is this false myth still propagated, in the face of overwhelming scientific evidence to the contrary? It's an interesting commentary on science and society, the media, and parents desperate to find the source of a heart-wrenching disease. The media has been blamed for significantly fanning the story, and continues to mention it. For example, there were five times the number of evening news stories on it in 2010 than in 2001, long after the science had thoroughly debunked it. Even celebrities got on the bandwagon—like Jenny McCarthy, who publicly blamed her son's autism on the MMR vaccine—which then spurred even more misinformation and Internet coverage. Many sites on the Internet also continue to promote this link (there are typically no referees or other quality controls on websites), and those opposed to vaccines continue to cite the original study.

A cornerstone to any scientific finding is the ability to be repeated by others and stand the test of time. The link between the MMR vaccine and autism does neither, and MMR continues to be a highly effective vaccine preventing diseases that, prior to the vaccine, maimed and killed countless children.

16: Bugs As Drugs

The Future

Sometime in the near future:

Doctor: Congratulations, the tests confirm that you're pregnant! Your fecal microbiota tests also suggest that there may be ways to improve the development of your baby during your pregnancy. Our nutritionist will suggest a modified diet that will alter your intestinal microbiota to enhance the health of your baby.

Doctor: Congratulations, it's going to be a girl! Everything looks great on the ultrasound, but your baby is still in a breech position and we might have to opt for a C-section if she doesn't turn around in time. One thing we may consider during a C-section is to take a vaginal swab before delivery and wipe your baby's mouth with it in order to give your baby the microbes you would have given her if she had been born vaginally. This will make her healthier in the long run.

Doctor: Your baby turned around just in time, and your vaginal delivery went great! As you know, you've been given antibiotics for Group B strep as a preventative measure, which may have altered your newborn's microbiota. Since you're breastfeeding, we'd like you to put a few drops of this solution on your breasts just prior to feeding. It contains a few probiotics that will replenish your baby's microbiota with microbes that are beneficial at this stage for your child's early mental and immune development.

Doctor: Don't worry, urinary tract infections like the one your daughter has are fairly common at six months of age. The antibiotic treatment worked nicely and the infection is now cleared. However, we noticed that because of the antibiotic we gave her, your child's gut is now missing some good bugs, putting her at risk for allergies and asthma. Here is a solution of four microbes that can be given orally that will replenish these organisms.

Doctor: Your toddler is developing beautifully! We noticed, however, that she has a certain microbe that might put her at increased risk for autism—nothing to worry too much about, especially because we can give her this pill that gets rid of this bad bug and leaves the good ones alone.

Doctor: Thanks for coming in for a three-year-old checkup! We noticed that when we tested your daughter's urine, there were certain molecules called metabolites that put her at a higher risk for inflammatory bowel disease. Luckily, we're going to give you this medicine that will help push her microbiota to a healthier population that will decrease her risk.

Doctor: I can't believe your daughter is starting kindergarten! It's great you're here for her vaccinations. We noticed she didn't have a really strong response to the measles vaccine we gave her when she was one, so we're going to give her this specially designed probiotic, which will help the booster shots work better.

Doctor: Thanks for bringing your daughter for her checkup before she starts first grade. She's doing fantastic, but we did notice that she has been gaining more weight than is ideal for her age. By having her wear the glucose monitor and doing an extensive analysis of the food she ate, we were able to determine which foods triggered a glucose spike based on her microbiota composition. The good news is that we're able to recommend a specific diet for her—this is her very own personalized diet, and yes, she can eat ice cream and pizza. However, there are some other foods that she should avoid, as they're the ones that trigger the weight gain.

Could these fictional conversations become reality in the future, indeed in a few short years? Definitely! All of the above examples are based on concepts that are already well under way in labs around the world, and are rapidly being developed and commercialized as new ways to promote health and prevent or treat diseases.

Understanding the Microbiome

The underlying concepts of the above conversations are that we can: a) rapidly determine a person's microbial composition, and b) actually do something to correct it if it needs altering. The first is relatively easy, and there are several companies that can analyze a person's microbiota for as little as $100 per sample. The trick is to know how to make sense of that analysis. The second concept is much more profound. Every person is born with the same DNA they die with—it is impossible to change our *Homo sapiens* DNA in our lifetime, other than perhaps through some skin cell mutations picked up from a bad sunburn in our youth (and even these changes won't alter the DNA that we pass on to our offspring). We do not evolve in our lifetime; it takes many generations to select and pass along genetic changes that then become part of the population.

However, we already know that we can rapidly shift the microbiota in and on our bodies—diet changes, antibiotics, probiotics, and fecal transfers all result in microbial changes. Specific gene therapy to correct a defective human gene is still an experimental technique, and unfortunately an early trial had a fatal outcome, which has slowed progress. But given that microbes have at least 150 times more genes than we do, and that we can readily change the microbes (and their genes), there is terrific excitement and promise in manipulating the microbiome in certain ways in children, and even in adults. You already do it every time you eat yogurt with probiotics in it, or when you travel to another country with a very different cuisine.

Research is getting closer to finding more effective ways of targeting and modifying our microbiome. This chapter explores some of these promising methods and their implications, and what is being done to make those conversations with the doctor a reality.

As DNA sequencing began to come on line in the early 1990s, we started to talk about sequencing the entire human genome. In the early days it was pie-in-the-sky discussion, much like the fictional idea from Jurassic Park (sequence a dinosaur's genome from a fossilized insect). However, the sequencing technology improved rapidly and, with a combined massive international effort, by 2003 we knew the entire genome sequence of humans. This was truly a remarkable scientific milestone (although to this day we still don't know exactly how many genes there really are). However, with the completion of the Human Genome Project, there were an awful lot of scientists with DNA sequencing machines wondering what they could sequence next. The microbiome represented a wonderfully appealing target because of its size—much bigger than the human genome and with lots more to sequence. This was only a few years ago, but at that time we knew very little about the composition of the microbiome; what we knew was solely based on our ability to grow a handful of microbes in the lab.

As a result the Human Microbiome Project was launched, with one of its major goals being to establish the composition and sequence of the human microbiome, much like the hugely successful human genome sequence. Samples from five different body sites (airways, intestines, mouth, skin, and vagina) were taken from more than a hundred "normal" people. What is "normal"? Young, healthy, no antibiotics—but of course there is still debate about what a normal human is. From this large and ambitious project we learned much of what we know about the human microbiota today, and hundreds of new bacterial genomes were sequenced. The results from this project came out in 2012, which is considered old news in this incredibly fast-paced field.

By the time the project began, scientists were starting to realize that the microbiota was important, and we generally assumed we would identify a core human microbiome that we all shared. Wrong! What we learned is that each person has his or her own set of microbes. This was spectacularly confusing (and frustrating), but it did keep the sequencers happy, as they had lots more to sequence. The finding that each individual has their own microbiota has held true, and for the vast majority of one's life it stays fairly constant. So how do we deal with the complexity of each person having a different microbiota? How can we come up with general microbiome therapies that work on most people if everyone is so different? Surprisingly, it isn't as impossible as it sounds.

A golden rule in biology is that if the function of something is important, it is used widely again and again (i.e., it is conserved). Based on this concept, if the microbiota carries out important functions, there must be something these microbes do that is in common. When scientists analyzed the microbiota not by the identity of its members but by what they do, the picture became much less confusing. Given that there's probably a common core of microbial genes that need to be turned on while living off a human, different microbes should have similar genes that do the same job. In other words, it doesn't really matter which microbe the gene is in, as long as it's there and its product is being made. When one looks at the microbiome from this functional point of view, there really does seem to be a set of core genes that are needed to make us the normal functioning human beings we are.

Analysis of *Your* Microbiome

There are two main methods for determining the composition of a person's microbiome. The first is to take a sample (i.e., feces) and sequence all the DNA in it; subtract out the human sequences and what's left are the microbial sequences. This is a labor-intensive (and expensive) way of doing it, and may be realistic for only a handful of people.

The alternative, and by far the more common way, is to only sequence a gene found in all bacteria called the 16S rRNA gene. Some parts of this gene are the same in all bacteria (this gives us a common handle to grab on to and is needed for sequencing), and other parts of the gene are different in different bacterial species, which gives a unique fingerprint for each microbe. The major advantage of this is that we don't have to grow the microbe in the lab (we still can't grow many of the microbes in the human body), and the amount of sequencing data obtained is a lot more manageable (by manageable, we mean being about half a million sequences per sample!). The companies offering to sequence your microbiota for a small fee are regularly doing this kind of sequencing—all you have to do is mail them a small fecal sample (lucky postman!).

The problem with both methods is that you need to figure out what this massive dump of data really means. This is where the science is these days: Many microbiologists are culturing fewer bacteria and behaving more like computer scientists, sitting in front of computer screens for a good part of the day. Bioinformatics, the science that uses computers to handle large biological data, plays a major part in this, as the data are extremely complex. Bioinformatic platforms need to be built, combining many different programs, with the output ideally telling you a) the composition of the microbiota in that sample, and more importantly b) what it means—is it good, bad, or do we just not know yet (yet is the operative word here). As discussed previously, because of the large differences between each person, this is actually quite difficult to do.

Beyond Genes: Microbial Metabolites

There's a third method of analysis that's developing rapidly, and will complement and possibly even replace the two DNA analyses mentioned above. Every microbe goes about its business, making small molecules (metabolites) as part of its normal life of breaking down food, generating energy, and just generally living. In the past decade science has made tremendous advances in analyzing small molecules using sophisticated machines called mass spectrometers ("mass spec"). These powerful machines can take a complex mixture of molecules and figure out the weight of each molecule in the mixture. Nearly every molecule has a unique weight, so this gives us a fingerprint of what's in the mix. The problem is that we need to know what molecule each mass represents, which is a problem if the mixture contains molecules that no one has seen before. So far, we can confidently detect about 20 percent of the human metabolites, and less than 1 percent of the microbial metabolites. However, this is where the action is, because these small molecules produced by microbes, by humans, or by both, are telling us how we interact with our microbes. Knowing the names of the microbes or even the names of their genes only tells us what these microbes *may* be doing. In contrast, information on all our metabolites, also known as metabolomics, tells us what the microbes are actually doing. This, in our view, is where the future is.

In one of the futuristic examples given above, the doctor mentioned analyzing a urine sample for metabolites, and based on that, figuring out what microbes might be in the gut—these are what we call disease biomarkers. In our work, we've found that we can identify metabolites in the urine from three-month-old children that are indicative of increased asthma risk years before a child actually gets the disease. Strikingly, some of the metabolites found in urine are produced by intestinal microbes. How did they get there? They travel. Although gut bacteria live in the gut, the molecules they make or modify can enter the human body and end up in the urine (or brain, or the placenta, or anywhere else in the body). This is how microbes talk to us, and how we listen to what they have to say. Luckily scientists are slowly getting better at listening to their signals—an area that could have a huge impact on diagnostics.

Although metabolomics is still in an early phase, as we figure out which metabolites are important and which bugs they come from, it will make a very powerful technique indeed to analyze our microbiota and how they relate to health or disease. In the not-so-distant future we'll be able to predict a child's risk for certain diseases before they occur, based on "the talk" from his or her microbiota.

Second Generation Probiotics

Okay, now that we can figure out your child's microbiome, and decide that perhaps it needs tweaking for optimal health, how do we go about it? As we said before, the good thing is that the microbiota is much easier to change than human genes. There are several methods already under way, with more sophisticated ones coming.

By far the oldest method is to use probiotics. These are live bacterial strains that you put in food or drink, that won't harm you, and that may or may not have any effect on you. We've been doing this ever since humans started eating fermented food such as sauerkraut centuries ago (microbes cause the "fermented" part of these foods). However, probiotic bacteria don't stick to an already full house in the gut, and can't easily become part of your microbiota. The solution? You have to take them daily in great numbers (10 billion plus), which is of course of great corporate benefit but not necessarily as effective as it could be.

There's an entire field dedicated to probiotics, with claims that they improve nearly everything you can imagine. The problem with the current field is twofold. First, it's usually only one strain that a company champions (e.g., certain *Lactobacillus* or *E. coli* strains), and it's hard to believe that a single, noncompetitive microbe can have so many profound health benefits when we know that the microbiota is such a complex ecological community. Second, probiotics are not currently regulated by the FDA, and do not have to go through the incredibly rigorous clinical trials that drugs do. This also means that beneficial claims made for a probiotic are not necessarily backed up by extensive clinical trials. Probiotics are also often designed for longer shelf life stability and ease of manufacturing, rather than for medicinal properties (since they aren't regulated). Of all of the dozens of probiotic options you see in a store, very few of them follow rigorous microbiological methods to ensure that the microbes will stay alive and active by the time you take them. However, as we have seen throughout this book, many hints and small studies indicate that probiotics have some beneficial effects. Ask your health practitioner to recommend probiotics that have been tested in clinical trials.

With the increasing knowledge of microbiota and how they work, second generation probiotics are going to play a major role in our health and disease fairly soon. Work is already under way to create even better probiotics, by using mixtures of several microbes together, instead of just one or two species, and by using microbes that are normal residents of the microbiota and have a documented beneficial role. This makes sense, as the aim is to deliver an entire functioning microbial community, which should probably colonize and work better than the current single strains.

Probiotics of the future will include microbes that are quite happy to colonize and remain within you. Current probiotics are also being altered to express anti-inflammatory products, or encode adhesins that promote their colonization in the gut. This has significant implications, as you wouldn't have to take them daily, but it will increase safety concerns and affect marketing. Whether these will be FDA regulated (to ensure safety and prove claims of efficacy) is a heated discussion these days.

Prebiotics

Another area that's seeing increased attention is the use of prebiotics. These are usually complex carbohydrates or sugars, such as fiber, that serve as a food source for certain types of microbes. The concept is that if you eat this specific microbe food, you will enhance those particular microbes.

Again, this concept has been around for a while, and has seen varying degrees of success. It's difficult to find a carbohydrate that enriches a single organism, so their effects tend to be broader. Like probiotics, prebiotics are not regulated, so verification of the claims in controlled clinical trials is usually lacking. However, with our increased knowledge of microbiota and the effects of different diets, it's not hard to imagine getting human volunteers to eat various sugars and follow specific diets, and then analyze their microbiota to define exactly how these prebiotics work. Diet changes alter microbiota, so if we can figure out exactly what changes different prebiotics cause, they could show much future promise. In the context of the microbiota, you truly are what you eat.

Back to the Future: Fecal Transfers

Throughout this book, we have discussed the concept that a child may have a microbiota composition that puts them at risk for IBD, obesity, asthma, or other disease, or one that has been altered and unbalanced by antibiotics, gut inflammation, etc. Significant inroads are currently being made to correct this. Fecal microbiota transfers (FMTs) have completely changed how we think about manipulating the microbiota. These involve the transfer of feces (and all the microbes that they contain) from one person to another, either orally or by enema. Although they have been used in China since the fourth century to treat diarrhea and other diseases, they have recently gained huge notoriety because a) in certain cases they work very well, and b) it is a very gross concept. However, for people who have long-suffered a debilitating disease, opting for an FMT is an easy choice.

Most people undergoing surgery are given antibiotics to prevent secondary infections. However, as we have discussed, antibiotics carpet bomb the microbiota, which can allow potentially harmful microbes to gain a foothold, especially if the person is old, sick, or otherwise more susceptible. One such bacterial pathogen seen under these conditions is *Clostridium difficile* or *C. diff*. This has caused major problems in hospitals all over Canada, the US, and elsewhere (see Poop vs. *C. Diff* on page 257 for a recent study on *C. diff* cases in children). The real problem is that antibiotics given to treat the *C. diff* infection are less than 20 percent effective (they actually caused the problem in the first place, so why would they work now?), and this is a deadly disease. Ironically, *C. diff* is a relatively wimpy pathogen, and is easily outcompeted by pretty much any normal microbe, which in a way explains why it doesn't usually cause infections in healthy people. Several studies have now shown that simple fecal transfers are over 90 percent effective to treat *C. diff* infections. The simple act of delivering fecal microbes, either through a nasal tube to the gut or by enema, cures a potentially fatal disease.

This is the true proof that microbiota manipulations have a real place in modern medicine. However, because fecal transfers are really body fluid transfers, concerns have been raised, and rightfully so. Remember the issues with blood donors and hepatitis C several years ago? The medical community didn't know about the hepatitis C virus then, so blood was not tested for it, resulting in many hepatitis cases after blood transfusions. The same happened with HIV in the 1980s. As a result, the FDA has now put very tight restrictions on fecal transfers, so that a physician or company has to do all the paperwork that is normally required for an investigational new drug, which amounts to roomfuls of documents. In the US, this has severely dampened the clinical use of fecal transfers. Because of its simplicity, there are now even YouTube videos showing how to do it yourself—don't! This is causing significant medical concern, as there are still risks associated with the process, especially in a non-clinical setting.

In addition, many experts on fecal transfers believe it's necessary to give a heavy dose of antibiotics before the transfer, in order to remove the unwanted microbiota and increase the chances of the donor microbiota sticking. Logically, this cannot be achieved if someone opts for the DIY method.

The spectacular success of fecal transfers for *C. diff* has unleashed a flurry of fecal transfer clinical trials for other microbiota-associated diseases, such as IBD and autism. Thus far the results have been mixed, and not nearly as successful as with *C. diff* infections. However, this makes sense. In *C. diff* cases, there is one known cause of the disease, so it really doesn't matter much which organisms you put in (i.e., whose feces you use), as long as *C. diff* gets booted out of your system. With IBD, it's a big ask of the incoming microbes—they have to be able to colonize an already inflamed gut (inflammation kills microbes), displace the current resident population, and then dampen the inflammation in a human genetically predisposed to this disease. Although we're starting to identify microbes that seem to be beneficial for IBD, it's important whose feces you use as a donor, and we just don't know enough about this yet.

RePOOPulating Our Gut

OK, feces are gross—enough already! There should be nicer ways to alter our microbes, right? There are certainly people working on this. Among them, Dr. Emma Allen-Vercoe, a collaborator of ours at the University of Guelph in Canada, has become very skilled at growing fecal microbes in fermenters (containers without air). Although a bit smelly (she has an entire floor dedicated to this for obvious reasons), she can grow defined cultures of twenty to thirty human microbiota species together in a fermenter, or as she calls them, a "robogut." Her team of scientists has even put this defined population into two people with *C. diff*, with the great advantage that it doesn't contain bodily fluids, nasty viruses, etc. In both of these cases the transplant worked well. Emma calls this concept rePOOPulating people, which is a wonderful term as far as we are concerned. Her team is now working at producing these microbes under special clean conditions so that they're pharmaceutical grade.

In addition, others are working at packing feces or microbial cultures into gelatin-coated capsules. You have to take a fair number of

these "pills," but they can be sugarcoated, and are not nearly as ob-
noxious as the alternative. There's no doubt that fecal transfers will
soon become a thing of the past as these more sophisticated methods
are further developed, but they have already served their purpose
by demonstrating how powerful a change in the microbiota can be.

Crystal Ball Time

Where are we going? We can now analyze our microbiota, but we
need intelligent ways of manipulating them. Commercial interest
in this area is exploding, with multimillion dollar deals being an-
nounced between pharmaceutical companies and smaller biotech
companies that are developing potential therapies. Some are working
on delivering specific microbial populations for defined benefits. We
know the microbiota has a profound influence early in life in determining how the immune system and the brain develop. This is an
area of huge potential therapy, and several companies are defining
groups of microbes that have direct effects on immune development.
They plan to use these as potential therapies for IBD, asthma, vaccine responses, and a multitude of other diseases.

More sophisticated methods are being developed to specifically
alter microbiota populations. In a recent example, a compound was
developed that specifically binds to a cavity-causing bacterium that
lives in the mouth. In addition to targeting that bacterium, the compound also had an antibacterial activity coupled to it so it would kill
only the microbes it bound. Using this technique one could specifically target a single microbe and remove it from the population,
another first in the microbiology field. This breakthrough suggests

that in diseases in which just one or a small number of microbes are associated with the disease, these microbes could be targeted without altering the rest of the microbial community, which is much different than the concept of antibiotics that cause huge collateral damage.

Another area that is being touted is "phage therapy." Bacteria, like us, have viruses that attack specific species. These are called bacteriophages. Because each phage targets a particular bacterial species, there's optimism that if viruses specific to a bad bug can be identified, produced, and delivered, they can be used to target that microbe in the body. It's a very appealing concept (let the phage do the hard work), but, like most things, has its problems. Phage therapy has been around for a long time (it was explored extensively in Russia to try to cure infections), but has not been integrated into Western medicine. The problem is that the bacteria don't really like to be blown apart by a phage (who does?), and quickly mutate to become resistant to the phage. Thus, if phage therapy is used extensively, we will see rapid resistance, just as we have seen with antibiotic resistance, rendering the phage therapy useless.

Recently there's been a major discovery of a system that can be used to target specific genes in most organisms, called CRISPR/Cas9. This system can be used to target a gene and cut it, killing the organism. Bacteria have used systems such as this as a type of immune system to defend against the viruses that infect them. Data are emerging to indicate we may be able to use this system to specifically target a unique microbe within the microbiota. This has obvious applications, but is still in its very early stages of development.

Personalized Diets

Let's finish discussing the future with what the microbiota does best—breaking down our food and making energy for us (and themselves). As we discussed in the chapter on obesity, microbiota play a role in this worldwide problem. As we all know, perhaps from personal experience, diets just don't work very well—we lose a bit of weight, but then we gain it back. Also, given what we now know, it's hard to believe that one diet would work for everyone, given the differences in our microbiota and what they do. We also know that some people seem to have all the luck—they can eat anything yet stay skinny as a rail, while most of us can only look on enviously. All this suggests that we should think about personalized diets tailored to our microbiota.

There's work already under way in Israel by Dr. Eran Elinav and his colleagues at the Weizmann Institute of Science, where they're working at personalizing an individual's diet based on their microbiome. They're using massive data analysis to correlate the microbiota and glucose (sugar) spikes in different people, and are finding, not surprisingly, that different individuals (depending on their microbiota) respond differently in their glucose spikes with different foods. They really did find people who could eat ice cream and pizza without causing glucose spikes (now that is one feces every kid should want!). Our presumption is that work such as this will completely change the entire dieting world, as we become much more sophisticated at designing personalized diets, with hopefully much better outcomes in controlling weight and other aspects of our health.

Given all this, we hope it's apparent that the fictional physician conversations at the beginning of this chapter are based on what's already being tested in labs around the world. This area is changing very fast, and the results will be in the clinic or on the market much faster than standard drugs for the reasons previously discussed. It's truly an exciting time, and we believe this will result in a major revolution in medicine, empowering us to deal with many of the most common health problems plaguing modern society in ways we couldn't even dream of a few short years ago. Hang on, it's going to be a fun ride!

Dos and Don'ts

◆ **Don't—** believe everything you hear about the microbiome; trust your physician to stay on top of what has been proven medically. It takes a long time from the eureka moment in the lab until something becomes common medical practice. There's an awful lot of information available that has no scientific backing, which makes things much more confusing. However, if a treatment passes full randomized clinical trials, and is FDA-approved, you can be certain it has been extensively tested, and the claims are real.

◆ **Do—** look into a fecal transfer, if you or anyone you know gets *C. diff* (but do *not* do it yourself!). Unfortunately, this is a common hospital infection, usually following surgery and antibiotics. The use of antibiotics to treat it has had a very poor success rate. There are now good clinical trials that prove that fecal transfers are much superior to antibiotics, even in kids, although there are still regulatory hurdles to go through as this becomes more mainstream.

◆ **Do—** pay attention: this is a rapidly changing field, and new treatments could come on line quickly, and might be of use for your child, perhaps even as part of a clinical trial. Because of the extensive testing needed for full clinical approval, one often has a pretty good idea if something works long before it is officially approved. By getting involved in clinical trials at an earlier stage, you can often benefit from these treatments before they're fully approved. If the treatment works spectacularly well, they will stop the trial before it's done, and even treat the controls. This happened for fecal transfers and *C. diff.* It was unethical to keep the controls untreated, as it was obvious it worked so well.

◆ **Don't—** believe that microbiota will cure everything—they can have effects, but these are part of complex interactions between complex populations of microbes with the environment and our genes. This is complex science, with many factors involved. Because of the complexity of the microbiota, and its overlapping functions, it's going to take a lot of science to figure out exactly how things work. However, there are more than enough examples to indicate that the microbiota plays a major role in both our health and disease. Right now it's an incredibly popular area of science, and there's a major bandwagon effect. As science plods on, it will tease out which effects are real and which ones might not be.

And finally, as we've said all along:

◆ **Do—** let your kid be a kid and interact with their world, and develop as kids have for the past million years. Let them eat dirt!

POOP VS. *C. DIFF*

C. diff infections in children are becoming more recurrent and severe. What's more alarming is that this infection is not just occurring in hospitals, but also in day cares and schools.

To address this, the Mayo Clinic in Rochester, Minnesota, began a fecal transplant program for children in 2013, something that had been avoided in the past due to safety concerns in dealing with pediatric patients. The results have been outstanding. Every one of the twenty-seven children treated so far dramatically improved almost immediately after the transplant (a 100 percent cure rate!). Many of the parents simply could not believe that a cure could occur so fast and so simply. Some parents came to this clinical study after their pediatricians told them that FMTs were dangerous, so it's crucial that more doctors are aware of these results and that more clinics start performing these transplants safely across North America.

Acknowledgments

First and foremost, we would like to thank our spouses and lifelong partners in parenting, Jane Finlay and Esteban Acuña. Their support, input, critical reading and editing, and help in so many other ways ("Can you get the kids . . . again?") has made this possible. We especially would like to thank Jane, who carefully read and edited all the chapters from the perspective of a practicing pediatrician and a certified infectious diseases expert to ensure the medical accuracy of the book's contexts and bring a real-life pediatrician point of view to it all, and to keep us PhDs on track.

So many people have been incredibly helpful in so many ways. Janis Sarra, Pieter Cullis, and Joel Bakan demystified the book agent process, and Joel introduced us to our wonderful literary agents, John Pearce and Chris Casuccio from Westwood Creative Artists. From the moment we sent our first pages to John and Chris, they were as committed to this project as we were. They have been a treat to work with and we are very grateful for all their intelligent, creative, and on-point suggestions; this book became a reality in great part thanks to them. More importantly, it was through them that we were able to work with Andra Miller from Algonquin Books and Nancy Flight from Greystone Books, our extremely talented editors who did so much to simplify this process for us, and provided such insightful, helpful guidance and editing of the entire book.

A wide range of people from numerous different areas of expertise helped us improve the manuscript, as this was our first attempt at writing a public book, rather than a scientific paper. We figured if they could understand it, we were on the right track. These include Nancy Gallini (an economist), Janis Sarra (a legal scholar), Toph Marshall (a Greek papyrus scholar!), Lara O'Donnell (a scientific program manager), Shaylih Muehlmann (an anthropologist and new mother who studies the Mexican drug trade while breastfeeding her son), Stefanie Vogt (a fellow microbiologist), Rocio Pazos (a high school teacher), Wendy Colling (a high school teacher), Mariela Podolski (a psychiatrist), Melania Acuña (a dentist), Shelly Blessin (an inventory analyst), Adriana Arrieta (an economist), Edmond Chan (a pediatrician), Stuart Turvey (a pediatrician-scientist), and Tobias Rees (an anthropologist).

So many other wonderfully generous people supplied personal and moving anecdotes, questions, and experiences, which are by far the most interesting part of the book. These include Pat Patrick, Ellen Bolte, Carley Akehurst, Thomas Louie, Emma Allen-Vercoe, Eran Elinav, Marjorie Harris, Julia Ewaschuk, Malcolm and Jennie Kendall, Veronica Niehaus, Margo Nelson, Joey Dubuc, Ivonne Montealegre, Anamaria Castillo, Astrid Antillón, Kristie Keeney, Jennifer Sweeten, Amanda Webster, Agnes Wong, Navkiran Gill, Amanda Roe, Erin Sawyer, Marilyn Robertson, Roxana Ramírez, and Fiorella Chinchilla.

As scientists, we have had the good fortune to have grant funding from the Canadian Institutes of Health Research (CIHR), Canadian Institute for Advanced Research (CIFAR), and other agencies that have allowed us to mentally and experimentally explore several of the topics discussed in this book. Brett would also like to thank the Peter Wall Institute for Advanced Studies for his endowed Chair and the Carnegie Foundation for funding his sabbatical stay in Scotland where much of this book was written, and much good whiskey was experimentally sampled (all in the name of science, of course).

After finishing this book we continue to be very interested in the topics at hand and welcome any questions or comments about what is covered in these pages. We have created www.letthemeatdirt.com, which provides a selection of practical information, as well as our contact information.

Progress Report:
Digging Up More Dirt

Seven thousand five hundred ninety. That is the number of scientific peer-reviewed studies that have been published on the human microbiome since we finished writing *Let Them Eat Dirt* only 14 months ago. The field is growing by 36 percent annually based on papers published. The human microbiome has captured the attention of thousands of scientists from many different disciplines worldwide, and its study is transforming the way we look at modern medicine and potential treatments for everything from obesity to cancer. Scientific conferences on subjects ranging from neurology to immunology to cancer now include at least one session dedicated to microbiome studies, and there are increasing numbers of microbiome-specific conferences held every year.

What's the upshot? We are adding to our knowledge at a rapid pace. Here's an overview of some of the most important recent developments:

Let's start with our bread and butter—asthma research. Our study of Canadian infants found that a reduction in four types of bacteria in babies could be detected months before asthma symptoms appeared, and that these four bacteria ameliorated the disease in mice. Armed with that data, we studied babies from a rural community in coastal Ecuador and found striking results. Not only were we able to detect microbiome changes in very young babies before they went on to develop asthma, we also found that the most profound differences involved fungi, not bacteria. Until recently, most microbiome studies surveyed only bacteria, but current methods now allow us to look for other microbes as well. Based on this work, it appears that fungi may be among the microbes that interact with the immune system early on, and they may also be involved in human asthma. Another study, recently published by Susan Lynch's lab at the University of California San Francisco, also found a strong association between fungal species and allergy risk, further suggesting that we must include fungi in our lab experiments if we want to get a better understanding of how asthma develops in children. Both of our labs (and likely Susan's, too) are now stinky with the moldy smell of fungi cultures, and we are eager to see what we find out next.

There is also exciting news from scientists who study the link between the gut microbiome and obesity. A cutting-edge study led by Israeli scientists Eran Segal and Eran Elinav found that the gut microbiome affects how our blood sugar levels respond to certain foods. For example, Claire's blood sugar might spike higher than Brett's after eating an orange, but the reverse effects may happen after we both eat a bowl of whole grain pasta. In large part, this is due to the differences in our microbiomes. The groundbreaking finding has led to a start-up company called DayTwo, which aims to provide personalized dietary recommendations based on an analysis of customers' individual microbiomes and blood glucose responses to food. This innovative approach—"personalized dieting"—involves completing an extensive survey of eating habits, obtaining a glucose blood test from a physician, and using a sample kit to collect and return a small poop sample for microbiome analysis.

IN ANOTHER STUDY, the Segal and Elinav research group found that, contrary to popular belief, artificial sweeteners lead to glucose intolerance, a pre-diabetic state consisting of high blood sugar. This occurs through the actions of the gut microbiome. This study was definitely not well received by the companies that make these sweeteners, nor by the millions of consumers who have been using artificial sweeteners for the exact opposite effect: to prevent weight gain, which is associated with glucose intolerance and diabetes. Skeptics claim that because the study was done in mice, which of course are very different from humans, it may not be valid. However, a recent study of over 3,000 mother-baby pairs in Canada has found a strong correlation between daily artificial sweetener use during pregnancy and an increase in body weight at one year of age, even after accounting for other obesity risk factors, such as maternal body weight, total caloric intake, and diet. While these results need to be confirmed by other studies in humans, the size and design of this one suggest its results are accurate, strongly discouraging the daily use of artificial sweeteners during pregnancy and childhood.

Yet another study from this research team is helping us understand how some people's microbiomes make it easier for them to gain weight and harder for them to lose it. Many of us diet after the holiday season, hoping to lose the weight we gained from all those decadent meals. Or we go on and off diets during the year, losing and regaining the same ten pounds. This kind of yo-yo dieting ultimately fails. In an intriguing study, again done with mice, Segal and Elinav found specific gut microbes that persist after successful dieting and weight loss, which later make it easier to regain the pounds. There is still a lot left to understand and the exact microbial culprits of that stubborn unwanted weight have yet to be found, but this is clearly an important development. We are following it closely—even as we drag ourselves to the gym!

Perhaps one of the most contentious issues we covered in *Let Them Eat Dirt* was the association between microbiome changes and Autism Spectrum Disorder (ASD). ASD attracts intense interest and response from the public, and our book was no exception. After the publication, we were approached by several scientists who disagreed with our decision to include ASD as a disorder associated with microbiome alterations. Their argument was that all published studies were done exclusively in animals, and that the human evidence we cited came from unpublished isolated cases treated with fecal transplant therapy. For these same reasons, we had discussed this issue at length during the writing process. We ultimately decided to include the research because of the quality of the animal studies published, as well as the direct communications we had with the physicians who had treated ASD children with fecal transplants.

More recently, another fecal transplant study was performed at Arizona State University. This study's findings support our views. In this clinical trial, 18 children diagnosed with ASD were first treated with the antibiotic vancomycin for 2 weeks, followed by a 12- to 24-hour fast with bowel cleansing and daily fecal transfers for 7 to 8 weeks. This study tracked gastrointestinal symptoms as well as behavioral symptoms related to ASD for 8 weeks post-treatment. It showed a drastic improvement in the constipation, diarrhea, indigestion, and abdominal pain that are common in ASD patients.

Astonishingly, clinical assessments also showed a significant improvement in the behavioral symptoms 8 weeks after treatment ended. Longer-term studies are clearly needed, but this important study still represents a sizable step forward in ASD therapy. As would be expected, larger-scale follow-up clinical trials are currently taking place. We encourage readers to follow these types of well-designed studies and to bring them to the attention of their health practitioners. These remain early days for this kind of therapy; it will take time—and more research—for it to become a valid mainstream treatment option.

What we *do* know is valid and accepted today is that microbes play a vital role in childhood health and development, and new studies will continue to explore their role. By embracing this new science, you can greatly improve your child's well-being. Love your microbes!

For Further Reading

Chapter 1

Cox, L. M., and M. J. Blaser. "Antibiotics in early life and obesity." *Nature Reviews Endocrinology* 11, no. 3 (2014).

Strachan, D. P. "Hay fever, hygiene, and household size." *BMJ* 299, no. 6710 (1989).

von Mutius, E. "Allergies, infections and the hygiene hypothesis—the epidemiological evidence." *Immunobiology* 212, no. 6 (2007).

Chapter 2

Arrieta, M. C., and B. B. Finlay. "The commensal microbiota drives immune homeostasis." *Frontiers in Immunology* 3 (2012).

Arrieta, M. C., L. T. Stiemsma, N. Amenyogbe, E. M. Brown, and B. B. Finlay. "The intestinal microbiome in early life: health and disease." *Frontiers in Immunolology* 5 (2014).

Clemente, J. C., L. K. Ursell, L. W. Parfrey, and R. Knight. "The impact of the gut microbiota on human health: an integrative view." *Cell* 148, no. 6 (2012).

Dominguez-Bello, M. G., M. J. Blaser, R. E. Ley, and R. Knight. "Development of the human gastrointestinal microbiota and insights from high-throughput sequencing." *Gastroenterology* 140, no. 6 (2011).

Sekirov, I., S. L. Russell, L. C. M. Antunes, and B. B. Finlay. "Gut microbiota in health and disease." *Physiological Reviews* 90, no. 3 (2010).

Wlodarska, M., A. D. Kostic, and R. J. Xavier. "An integrative view of microbiome-host interactions in inflammatory bowel diseases." *Cell Host & Microbe* 17, no. 5 (2015).

Wrangham, R., and R. Carmody. "Human adaptation to the control of fire." *Evolutionary Anthropology* 19, no. 5 (2010).

Chapter 3

Aagaard, K., J. Ma, K. M. Antony, R. Ganu, J. Petrosino, and J. Versalovic. "The placenta harbors a unique microbiome." *Science Translational Medicine* 6, no. 237 (2014)5.

Aagaard, K., K. Riehle, J. Ma, N. Segata, T. A. Mistretta, C. Coarfa, S. Raza, S. Rosenbaum, I. van den Veyver, A. Milosavljevic, D. Gevers, C. Huttenhower, J. Petrosino, and J. Versalovic. "A metagenomic approach to characterization of the vaginal microbiome signature in pregnancy." *PLOS ONE* 7, no. 6 (2012).

Jašarević, E., C. L. Howerton, C. D. Howard, and T. L. Bale. "Alterations in the Vaginal Microbiome by Maternal Stress Are Associated With Metabolic Reprogramming of the Offspring Gut and Brain." *Endocrinology* 156, no. 9 (2015).

Koren, O., J. K. Goodrich, T. C. Cullender, A. Spor, K. Laitinen, H. Kling Bäckhed, A. Gonzalez, J. Werner, L. Angenent, R. Knight, F. Bäckhed, E. Isolauri, S. Salminen, and R. Ley. "Host remodeling of the gut microbiome and metabolic changes during pregnancy." *Cell* 150, no. 3 (2012).

Zijlmans, M. A. C., K. Korpela, J. M. Riksen-Walraven, W. M. de Vos, and C. de Weerth. "Maternal prenatal stress is associated with the infant intestinal microbiota." *Psychoneuroendocrinology* 53 (2015).

Chapter 4

Dominguez-Bello, M. G., E. K. Costello, M. Contreras, M. Magris, G. Hidalgo, N. Fierer, and R. Knight. "Delivery mode shapes the acquisition and structure of the initial microbiota across multiple body habitats in newborns." *Proceedings of the National Academy Sciences* 107, no. 26 (2010).

Lotz, M., D. Gütle, S. Walther, S. Ménard, C. Bogdan, and M. W. Homef. "Postnatal acquisition of endotoxin tolerance in intestinal epithelial cells." *Journal of Experimental Medicine* 203, no. 4 (2006).

Neu, J., and J. Rushing. "Cesarean versus vaginal delivery: long-term infant outcomes and the hygiene hypothesis." *Clinics in Perinatology* 38, no. 2 (2011).

Thanabalasuriar, A., and P. Kubes. "Neonates, Antibiotics and the Microbiome." *Nature Medicine* 20, no. 5 (2014).

Thysen, A. H., J. M. Larsen, M. A. Rasmussen, J. Stokholm, K. Bønnelykke, H. Bisgaard, and S. Brix. "Prelabor cesarean section bypasses natural immune cell maturation." *Journal of Allergy and Clinical Immunology* 136, no. 4 (2015).

Chapter 5

Arroyo, R., V. Martín, A. Maldonado, E. Jiménez, L. Fernández, and J. M. Rodríguez. "Treatment of infectious mastitis during lactation: antibiotics versus oral administration of Lactobacilli isolated from breast milk." *Clinical Infectious Diseases* 50, no. 12 (2010).

Bäckhed, F., J. Roswall, Y. Peng, Q. Feng, H. Jia, P. Kovatcheva-Datchary,

Y. Li, Y. Xia, H. Xie, H. Zhong, M. Khan, J. Zhang, J. Li, L. Xiao, J. Al-Aama, D. Zhang, Y. Lee, D. Kotowska, C. Colding, V. Tremaroli, Y. Yin, S. Bergman, X. Xu, L. Madsen, K. Kristiansen, J. Dahlgren, and J. Wang. "Dynamics and Stabilization of the Human Gut Microbiome during the First Year of Life." *Cell Host & Microbe* 17, no. 5 (2015).

Cabrera-Rubio, R., M. C. Collado, K. Laitinen, S. Salminen, E. Isolauri, and A. Mira. "The human milk microbiome changes over lactation and is shaped by maternal weight and mode of delivery." *American Journal of Clinical Nutrition* 96, no. 3 (2012).

Nylund, L., R. Satokari, S. Salminen, and W. M. de Vos. "Intestinal Microbiota During Early Life—Impact on Health and Disease." *Proceedings of the Nutrition Society.* 73, no. 4 (2014).

Rautava, S., R. Luoto, S. Salminen, and E. Isolauri. "Microbial contact during pregnancy, intestinal colonization and human disease." *Nature Reviews Gastroenterology & Hepatology* 9, no. 10 (2012).

Chapter 6

Du Toit, G., G. Roberts, P. H. Sayre, H. T. Bahnson, S. Radulovic, A. F. Santos, H. A. Brough, D. Phippard, M. Basting, M. Feeney, V. Turcanu, M. L. Sever, M. Gomez-Lorenzo, M. Plaut, and G. Lack. "Randomized trial of peanut consumption in infants at risk for peanut allergy." *New England Journal of Medicine* 372, no. 9 (2015).

Krebs, N. F., and K. M. Hambidge. "Complementary feeding: clinically relevant factors affecting timing and composition." *American Journal of Clinical Nutrition* 85, no. 2 (2007).

Parnell, J. A., and R. A. Reimer. "Prebiotic fiber modulation of the gut microbiota improves risk factors for obesity and the metabolic syndrome." *Gut Microbes* 3, no. 1 (2012).

Prescott, S., and A. Nowak-Węgrzyn. "Strategies to prevent or reduce allergic disease." *Annals of Nutrition and Metabolism* 59, suppl. 1 (2011).

Sonnenburg, E. D., and J. L. Sonnenburg. "Starving our microbial self: the deleterious consequences of a diet deficient in microbiota-accessible carbohydrates." *Cell Metabolism* 20, no. 5 (2014).

Chapter 7

Dellit, T. H., R. C. Owens, J. E. McGowan, D. N. Gerding, R. A. Weinstein, J. P. Burke, W. C. Huskins, D. L. Paterson, N. O. Fishman, C. F. Carpenter, P. J. Brennan, M. Billeter, and T. M. Hooton. "Infectious Diseases Society of America and the Society for Healthcare Epidemiology of America guidelines for developing an institutional program to enhance antimicrobial stewardship." *Clinical Infectious Diseases* 44, no. 2 (2007).

Dethlefsen, L., S. Huse, M. L. Sogin, and D. A. Relma. "The pervasive effects of an antibiotic on the human gut microbiota, as revealed by deep 16S rRNA sequencing." *PLOS Biology* 6, no. 11 (2008).

Hempel, S., S. J. Newberry, A. R. Maher, Z. Wang, J. N. V. Miles, R. Shanman, B. Johnsen, and P. G. Shekelle. "Probiotics for the Prevention and Treatment of Antibiotic-Associated Diarrhea: A Systematic Review and Meta-analysis." *JAMA* 307, no. 18 (2012).

Jernberg, C., S. Lofmark, C. Edlund, and J. K. Jansson. "Long-term impacts of antibiotic exposure on the human intestinal microbiota." *Microbiology* 156, pt. 11 (2010).

Marra, F., C. A. Marra, K. Richardson, L. D. Lynd, A. Kozyrskyj, D. M. Patrick, W. R. Bowie, and J. M. FitzGerald. "Antibiotic use in children is associated with increased risk of asthma." *Pediatrics* 123, no. 3 (2009).

Van Boeckel, T. P., C. Brower, M. Gilbert, B. T. Grenfell, S. A. Levin, T. P. Robinson, A. Teillant, and R. Laxminarayan. "Global trends in antimicrobial use in food animals." *Proceedings of the National Academy of Sciences* 112, no. 18 (2015).

Van Boeckel, T. P., S. Gandra, A. Ashok, Q. Caudron, B. T. Grenfell,

S. A. Levin, and R. Laxminarayan. "Global antibiotic consumption 2000 to 2010: an analysis of national pharmaceutical sales data." *The Lancet Infectious Diseases* 14, no. 8 (2014).

Chapter 8

Azad, M. B., T. Konya, H. Maughan, D. S. Guttman, C. J. Field, M. R. Sears, A. B. Becker, J. A. Scott, and A. L. Kozyrskyj. "Infant gut microbiota and the hygiene hypothesis of allergic disease: impact of household pets and siblings on microbiota composition and diversity." *Allergy, Asthma & Clinical Immunology* 9, no. 1 (2013).

Fujimura, K. E., T. Demoor, M. Rauch, A. A. Faruqi, S. Jang, C. C. Johnson, H. A. Boushey, E. Zoratti, D. Ownby, N. W. Lukacs, and S. V. Lynch. "House dust exposure mediates gut microbiome Lactobacillus enrichment and airway immune defense against allergens and virus infection." *Proceedings of the National Academy of Sciences* 111, no. 2 (2013).

Pelucchi, C., C. Galeone, J. Bach, C. La Vecchia, and L. Chatenoud. "Pet exposure and risk of atopic dermatitis at the pediatric age: A meta-analysis of birth cohort studies." *Journal of Allergy and Clinical Immunology* 132 no. 3 (2013).

Chapter 9

Cherednichenko, G., R. Zhang, R. A. Bannister, V. Timofeyev, N. Li, E. B. Fritsch, W. Feng, G. C. Barrientos, N. H. Schebb, B. D. Hammock, K. G. Beam, N. Chiamvimonvat, and I. N. Pessah. "Triclosan impairs excitation-contraction coupling and Ca2+ dynamics in striated muscle." *Proceedings of the National Academy of Sciences* 109, no. 35 (2012).

Hesselmar, B., A. Hicke-Roberts, and G. Wennergren, "Allergy in Children in Hand Versus Machine Dishwashing." *Pediatrics* 135, no. 3 (2015).

Hesselmar, B., F. Sjoberg, R. Saalman, N. Aberg, I. Adlerberth, and A. E. Wold. "Pacifier Cleaning Practices and Risk of Allergy Development." *Pediatrics* 131, no. 6 (2013).

Tung, J., L. B. Barreiro, M. B. Burns, J. Grenier, J. Lynch, L. E. Grieneisen, J. Altmann, S. C. Alberts, R. Blekhman, and E. A. Archie. "Social networks predict gut microbiome composition in wild baboons." *Elife* 4 (2015).

Chapter 10

Cox, L. M., S. Yamanishi, J. Sohn, A. V. Alekseyenko, J. M. Leung, I. Cho, S. G. Kim, H. Li, Z. Gao, D. Mahana, J. Zárate Rodriguez, A. Rogers, N. Robine, P. Loke, and M. Blaser. "Altering the Intestinal Microbiota during a Critical Developmental Window Has Lasting Metabolic Consequences." *Cell* 158, no. 4 (2014).

Kleiman, S. C., H. J. Watson, E. C. Bulik-Sullivan, E. Y. Huh, L. M. Tarantino, C. M. Bulik, and I. M. Carroll. "The Intestinal Microbiota in Acute Anorexia Nervosa and During Renourishment." *Psychosomatic Medicine* 77, no. 9 (2015).

Magrone, T., and E. Jirillo. "Childhood Obesity: Immune Response and Nutritional Approaches." *Frontiers in Immunology* 6 (2015).

Park, S., and J. H. Bae. "Probiotics for weight loss: a systematic review and meta-analysis." *Nutrition Research* 35, no. 7 (2015).

Turnbaugh, P. J., F. Bäckhed, L. Fulton, and J. I. Gordon. "Diet-Induced Obesity Is Linked to Marked but Reversible Alterations in the Mouse Distal Gut Microbiome." *Cell Host & Microbe* 3, no. 4 (2008).

Turnbaugh, P. J., R. E. Ley, M. A. Mahowald, V. Magrini, E. R. Mardis, and J. I. Gordon. "An obesity-associated gut microbiome with increased capacity for energy harvest." *Nature* 444, no. 7122 (2006).

Chapter 11

Hartstra, A. V., K. E. C. Bouter, F. Bäckhed, and M. Nieuwdorp. "Insights Into the Role of the Microbiome in Obesity and Type 2 Diabetes." *Diabetes Care* 38, no. 1 (2015).

Hu, C., F. S. Wong, and L. Wen, "Type 1 diabetes and gut microbiota: Friend or foe?" *Pharmacological Research* 98 (2015).

Karlsson, F. H., V. Tremaroli, I. Nookaew, G. Bergström, C. J. Behre, B. Fagerberg, J. Nielsen, and F. Bäckhed., "Gut metagenome in European women with normal, impaired and diabetic glucose control." *Nature* 498, no. 7452 (2013).

Qin, J., Y. Li, Z. Cai, S. Li, J. Zhu, F. Zhang, S. Liang, W. Zhang, Y. Guan, D. Shen, Y. Peng, D. Zhang, Z. Jie, W. Wu, Y. Qin, W. Xue, J. Li, L. Han, D. Lu, P. Wu, Y. Dai, X. Sun, Z. Li, A. Tang, S. Zhong, X. Li, W. Chen, R. Xu, M. Wang, Q. Feng, M. Gong, J. Yu, Y. Zhang, M. Zhang, T. Hansen, G. Sanchez, J. Raes, G. Falony, S. Okuda, M. Almeida, E. LeChatelier, P. Renault, N. Pons, J. Batto, Z. Zhang, H. Chen, R. Yang, W. Zheng, S. Li, H. Yang, J. Wang, S. D. Ehrlich, R. Nielsen, O. Pedersen, K. Kristiansen, and J. Wang. "A metagenome-wide association study of gut microbiota in type 2 diabetes." *Nature* 490, no. 7418 (2012).

Vrieze, A., E. van Nood, F. Holleman, J. Salojärvi, R. S. Kootte, J. F. W. M. Bartelsman, G. M. Dallinga-Thie, M. T. Ackermans, M. J. Serlie, R. Oozeer, M. Derrien, A. Druesne, J. E. van Hylckama Vlieg, V. W. Bloks, A. K. Groen, H. G. Heilig, E. G. Zoetendal, E. S. Stroes, W. M. de Vos, J. B. Hoekstra, and M. Nieuwdorp. "Transfer of Intestinal Microbiota from Lean Donors Increases Insulin Sensitivity in Individuals with Metabolic Syndrome." *Gastroenterology* 143, no. 4 (2012).

Chapter 12

Collins, S. M. "A role for the gut microbiota in IBS." *Nature Reviews Gastroenterology & Hepatology* 11, no. 8 (2014).

Colman, R. J., and D. T. Rubin. "Fecal microbiota transplantation as therapy for inflammatory bowel disease: A systematic review and meta-analysis." *Journal of Crohn's and Colitis* 8, no. 12 (2014).

de Sousa Moraes, L. F., L. M. Grzeskowiak, T. F. de Sales Teixeira, and M. D. C. Gouveia Peluzio. "Intestinal Microbiota and Probiotics in Celiac Disease." Clinical Microbiology Reviews 27, no. 3 (2014).

de Weerth, C., S. Fuentes, and W. M. de Vos. "Crying in infants: on the possible role of intestinal microbiota in the development of colic." *Gut Microbes* 4, no. 5 (2013).

de Weerth, C., S. Fuentes, P. Puylaert, and W. M. de Vos. "Intestinal Microbiota of Infants with Colic: Development and Specific Signatures." *PEDIATRICS* 131, no. 2 (2013).

Gevers, D., S. Kugathasan, L. A. Denson, Y. Vázquez-Baeza, W. Van Treuren, B. Ren, E. Schwager, D. Knights, S. Song, M. Yassour, X. Morgan, A. Kostic, C. Luo, A. González, D. McDonald, Y. Haberman, T. Walters, S. Baker, J. Rosh, M. Stephens, M. Heyman, J. Markowitz, R. Baldassano, A. Griffiths, F. Sylvester, D. Mack, S. Kim, W. Crandall, J. Hyams, C. Huttenhower, R. Knight, and R. Xavier. "The Treatment-Naive Microbiome in New-Onset Crohn's Disease." *Cell Host & Microbe* 15, no. 3 (2014).

Wacklin, P., P. Laurikka, K. Lindfors, P. Collin, T. Salmi, M. Lähdeaho, P. Saavalainen, M. Mäki, J. Mättö, K. Kurppa, and K. Kaukinen. "Altered Duodenal Microbiota Composition in Celiac Disease Patients Suffering from Persistent Symptoms on a Long-Term Gluten-Free Diet." *American Journal of Gastroenterology* 109, no. 12 (2014).

Chapter 13

Arrieta, M. C., and B. Finlay. "The intestinal microbiota and allergic asthma." *Journal of Infection* 69, suppl. 1 (2014).

Arrieta, M. C., L. T. Stiemsma, P. A. Dimitriu, L. Thorson, S. Russell, S. Yurist-Doutsch, B. Kuzeljevic, M. J. Gold, H. M. Britton, D. L. Lefebvre, P. Subbarao, P. Mandhane, A. Becker, K. M. McNagny, M. R. Sears, T. Kollmann, W. W. Mohn, S. E. Turvey, and B. Finlay. "Early infancy microbial and metabolic alterations affect risk of childhood asthma." *Science Translational Medicine* 7, no. 307 (2015).

Holbreich, M., M. Stein, R. Anderson, N. Metwali, P. S. Thorne, D. Vercelli, E. von Mutius, and C. Ober. "Allergic Sensitization and Enviromental Exposures in Amish and Hutterite Children." *Journal of Allergy and Clinical Immunology* 133, no. 2 (2014).

Ly, N. P., A. Litonjua, D. R. Gold, and J. C. Celedón. "Gut microbiota, probiotics, and vitamin D: Interrelated exposures influencing allergy, asthma, and obesity?" *Journal of Allergy and Clinical Immunology* 127, no. 5 (2011).

Olszak, T., D. An, S. Zeissig, M. P. Vera, J. Richter, A. Franke, J. N. Glickman, R. Siebert, R. M. Baron, D. L. Kasper, and R. S. Blumberg. "Microbial Exposure During Early Life Has Persistent Effects on Natural Killer T Cell Function." *Science* 336, no. 6080 (2012).

Russell, S. L., M. J. Gold, M. Hartmann, B. P. Willing, L. Thorson, M. Wlodarska, N. Gill, M. Blanchet, W. W. Mohn, K. M. McNagny, and B. B. Finlay. "Early life antibiotic-driven changes in microbiota enhance susceptibility to allergic asthma." *EMBO Reports* 13, no. 5 (2012).

Schaub, B., J. Liu, S. Höppler, I. Schleich, J. Huehn, S. Olek, G. Wieczorek, S. Illi, and E. von Mutius. "Maternal farm exposure modulates neonatal immune mechanisms through regulatory T cells." *Journal of Allergy and Clinical Immunology* 123, no. 4 (2009).

Sharief, S., S. Jariwala, J. Kumar, P. Muntner, and M. L. Melamed.

"Vitamin D levels and food and environmental allergies in the United States: Results from the National Health and Nutrition Examination Survey 2005–2006." *Journal of Allergy and Clinical Immunology* 127, no. 5 (2011).

Chapter 14

Braniste, V., M. Al-Asmakh, C. Kowal, F. Anuar, A. Abbaspour, M. Toth, A. Korecka, N. Bakocevic, L. G. Ng, P. Kundu, B. Gulyas, C. Halldin, K. Hultenby, H. Nilsson, H. Hebert, B. T. Volpe, B. Diamond, and S. Pettersson. "The gut microbiota influences blood-brain barrier permeability in mice." *Science Translational Medicine* 6, no. 263 (2014).

Hsiao, E. Y., S. W. McBride, S. Hsien, G. Sharon, E. R. Hyde, T. McCue, J. A. Codelli, J. Chow, S. Reisman, J. Petrosino, P. Patterson, and S. Mazmanian. "Microbiota modulate behavioral and physiological abnormalities associated with neurodevelopmental disorders." *Cell* 155, no. 7 (2013).

Pärtty, A., M. Kalliomäki, P. Wacklin, S. Salminen, and E. Isolauri. "A possible link between early probiotic intervention and the risk of neuropsychiatric disorders later in childhood: a randomized trial." *Pediatric Research* 77, no. 6 (2015).

Petra, A. I., S. Panagiotidou, E. Hatziagelaki, J. M. Stewart, P. Conti, and T. C. Theoharides. "Gut-Microbiota-Brain Axis and Its Effect on Neuropsychiatric Disorders with Suspected Immune Dysregulation." *Clinical Therapeutics* 37, no. 5 (2015.

Chapter 15

Ang, L., S. Arboleya, G. Lihua, Y. Chuihui, Q. Nan, M. Suarez, G. Solís, C. G. de los Reyes-Gavilán, and M. Gueimonde. "The establishment of the infant intestinal microbiome is not affected by rotavirus vaccination." *Scientific Reports* 4 (2014).

Kimmel, S. R. "Vaccine Adverse Events: Separating Myth from Reality." *American Family Physician* 66, no. 11 (2002).

Valdez, Y., E. M. Brown, and B. B. Finlay. "Influence of the microbiota on vaccine effectiveness." *Trends in Immunology* 35, no. 11 (2014).

Chapter 16

Zeevi, D., T. Korem, N. Zmora, D. Israeli, D. Rothschild, A. Weinberger, O. Ben-Yacov, D. Lador, T. Avnit-Sagi, M. Lotan-Pompan, J. Suez, J. A. Mahdi, E. Matot, G. Malka, N. Kosower, M. Rein, G. Zilberman-Schapira, L. Dohnalová, M. Pevsner-Fischer, R. Bikovsky, Z. Halpern, E. Elinav, and E. Segal. "Personalized Nutrition by Prediction of Glycemic Responses." *Cell* 163, no. 5 (2015).

Progress Report: Digging Up More Dirt

Azad, M. B. et al. "Association Between Artificially Sweetened Beverage Consumption During Pregnancy and Infant Body Mass Index." *JAMA Pediatrics* 170, no. 7 (2016).

Fujimura, K. E. et al. "Neonatal Gut Microbiota Associates with Childhood Multisensitized Atopy and T Cell Differentiation." *Nature Medicine* 22, no. 10 (2016).

Kang, D. W. et al. "Microbiota Transfer Therapy Alters Gut Ecosystem and Improves Gastrointestinal and Autism Symptoms: an Open-Label Study." *Microbiome* 5, no. 1 (2017).

Suez, J. et al. "Artificial Sweeteners Induce Glucose Intolerance by Altering the Gut Microbiota." *Nature* 514, no. 7521 (2014).

Thaiss, C. A. et al. "Persistent Microbiome Alterations Modulate the Rate of Post-Dieting Weight Regain." *Nature* 540, no. 7634 (2016).

Zeevi, D. et al. "Personalized Nutrition by Prediction of Glycemic Responses." *Cell* 163, no. 5 (2015).

Index

in infant formulas, 80, 81
irritable bowel syndrome treatment, 182
mastitis treatment, 78–79
during pregnancy, 39–40, 46, 48, 60, 164, 200
for premature babies, 67
type 2 diabetes management, 169
weight control, 153
psychiatric and psychological disorders. *See* brain function and neurological disorders; stress, depression, and anxiety
psychobiotics, 220–21

rabies, 204

sandboxes, 137–38
seeding with vaginal secretions, 61–63
short-chain fatty acids (SCFA), 28–29, 170
16S analysis, 243–44
soaps and sanitizers, 130–31
solid foods
    allergenic foods, 94–97
    in conjunction with breastfeeding or formula, 93–94, 97–98
    cultural preferences, 100
    dos and don'ts, 98–99
    fermented foods, 94
    fostering of good eating habits, 91, 142–43
    gluten in, 180

promotion of microbial diversity, 87–91, 141–43
    when and how to introduce, 91–94, 96–97, 180
Strachan, David, 6–7
stress, depression, and anxiety
    correlation with anorexia, 158
    early-life stress, 211
    irritable bowel syndrome risk, 181, 209
    microbiota changes and, 158–59, 205–8, 211–12
    during pregnancy, 40–42, 211, 213
    probiotic treatment, 211, 220–21
superbugs, 106

tantrums, 209–10
Toxoplasma parasite, 205
toys, washing of, 136–37
triclosan in hygiene products, 131
type 1 diabetes, 162, 165–67
type 2 diabetes
    antibiotics and, 110
    in children, 167
    correlation with obesity, 162–63, 167–68
    fecal microbiota transfer for, 168–69
    metformin treatment, 169, 170
    microbiota and, 89, 167–69, 170
    risk following gestational diabetes, 163, 167

ulcerative colitis. *See* inflammatory bowel diseases (IBD)

CARLOS TAYLHARDAT

CARLOS TAYLHARDAT

B. Brett Finlay, PhD, is the Peter Wall Distinguished Professor at the University of British Columbia and a world leader on the workings of bacterial infections. He has studied microbes for more than thirty years and has published more than 450 scientific articles. A cofounder of the biotech companies Inimex, Vedanta, and Microbiome Insights, Brett is Officer of the Order of Canada, the highest Canadian civilian recognition. He lives in Vancouver, BC, with his wife, who is a pediatrician, and has two grown children.

Marie-Claire Arrieta, PhD, is an assistant professor at the University of Calgary. Her recent study connecting asthma in very young babies to missing key intestinal bacterial species was deemed a breakthrough in the field and was reported by news outlets around the world in 2015. Arrieta has been published in leading scientific journals such as *Gastroenterology*, *PNAS*, and *Science Translational Medicine*. She spends her busy days juggling experiments, science writing, and play dates for her two young children.